Alison Richard is the Crosby Professor of ‌
ment Emerita and Senior Research Scient‌
She previously served as Vice Chancellor ‌
Cambridge, and in 2010 she was awarded ‌
mander of the British Empire) for her services to higher education.

Praise for *The Sloth Lemur's Song*

'Full of wonder and forensic intelligence, *The Sloth Lemur's Song* is a love song to the astonishing evolution of Madagascar. It is a fascinating journey from the island's origins to the complex tensions of the present day, with Alison Richard the most considerate and engaging of guides.'   Isabella Tree, bestselling author of *Wilding*

'This book is an encyclopedia of wonders, but it's also a riveting story of evolution through time in a land utterly unique. Madagascar is arguably the most amazing place on Earth. Richard knows it as few outsiders ever will, and its praises have never been better sung.'

David Quammen, bestselling author of *Spillover*

'Truly mind-blowingly epic . . . For every adventure you need a perceptive, intelligent and compassionate guide. Ours is author Alison Richard whose life's work has been Madagascar . . . A capitalise tale of enchanting and endangered biodiversity'

Resurgence and Ecologist magazine

'Brilliant . . . This is simply a wonderful book. Richard tells Madagascar's often improbable history with vivid detail and personal story based on her research, all backed up with the latest scientific thinking . . . You will enjoy the stories so much you may not notice that your world is expanding.'   Cool Green Science blog

'A love story; an ode to Madagascar. Throughout, the author interweaves first-person accounts of her extensive experience as a field biologist, detailed and accurate accounts of the natural history of the island, up-to-the-minute summaries of the latest scientific studies spanning everything from botany to geology to climatology, with the binding "through line" of the Malagasy people and their relationship to the landscape.'

Anne Yoder, Duke University

'With deep reflections about a culture immersed in bountiful nature, Richard guides the reader through Madagascar's transformations over millions of years . . . [and] reminds us what was lost, what remains, and what is under threat – unless we act.'

Yolanda Kakabadse, World Wildlife Fund International

'Richard shares her long experience, deep knowledge, and sincere and uncompromising passion for Madagascar, its people, and the history of its biodiversity. *The Sloth Lemur's Song* is essential reading for anyone who is passionate about Madagascar's people and nature, curious about its past and interested in its future.'

Olivier Langrand, executive director,
Critical Ecosystem Partnership Fund

'This remarkable new book is a captivating and absorbing account. Richard shows the importance of partnerships between all stakeholders, from the local to the international level, for sustainable biodiversity conservation.'

Jonah H. Ratsimbazafy, president,
Groupe d'Etude et de Recherche sur les
Primates de Madagascar (GERP)

THE SLOTH LEMUR'S SONG

Madagascar from the Deep Past
to the Uncertain Present

Alison Richard

The University of Chicago Press

The University of Chicago Press, Chicago 60637
© 2022 by Alison Richard
Alison Richard asserts the moral right to be acknowledged as the author of this work.
All rights reserved. No part of this book may be used or reproduced in any manner
whatsoever without written permission, except in the case of brief quotations in critical
articles and reviews. For more information, contact the University of Chicago Press,
1427 E. 60th St., Chicago, IL 60637.
Published 2022
Paperback edition 2023
Printed in the United States of America

32 31 30 29 28 27 26 25 24 23   1 2 3 4 5

ISBN-13: 978-0-226-81756-9 (cloth)
ISBN-13: 978-0-226-82949-4 (paper)
ISBN-13: 978-0-226-81757-6 (e-book)
DOI: https://doi.org/10.7208/chicago/9780226817576.001.0001

Originally published in the English language by HarperCollins Publishers Ltd.

LIBRARY OF CONGRESS CATALOGING-IN-PUBLICATION DATA

Names: Richard, Alison F., author.
Title: The sloth lemur's song : Madagascar from the deep past to the uncertain present
    / Alison Richard.
Description: Chicago : University of Chicago Press, 2022. | Includes bibliographical
    references and index.
Identifiers: LCCN 2021036595 | ISBN 9780226817569 (cloth) | ISBN
    9780226817576 (ebook)
Subjects: LCSH: Natural history—Madagascar.
Classification: LCC QH21.M28 R53 2022 | DDC 508.691—dc23
LC record available at https://lccn.loc.gov/2021036595

♾ This paper meets the requirements of ANSI/NISO Z39.48-1992 (Permanence
of Paper).

For Bessie, Charlotte, Alex and Ed

'Hope is not optimism, which expects things to turn out well, but something rooted in the conviction that there is good worth working for.'

Vaclav Havel, as recounted by Seamus Heaney

CONTENTS

PREFACE

It was the autumn of 1968. I was an undergraduate and aspiring anthropologist at Cambridge University, and I was at a lecture about the lemurs of Madagascar. They were interesting animals, to be sure, but the slides of the southern forests were what transfixed me. Silvery, spine-studded trees snaked toward the sky, and graceful, slender-trunked trees lofted flowers at the tips of silvery branches. I had never seen a silver forest before, and was instantly hooked. I have been 'living Madagascar' ever since.

This book is about what I have learned. It begins with a nondescript patch of ground in the middle of a huge supercontinent 250 million years ago. The patch drifts south with the supercontinent, eventually breaks loose, and becomes the island we call Madagascar. The land is home to a parade of animals during these epochs. Its most ancient occupants give way to lumbering dinosaurs and furry little mammals, and then they too disappear after an asteroid hits Earth 66 million years ago. The deep, wide sea now surrounding Madagascar makes it difficult to reach, but a handful of animals beat the odds and settle there. Their descendants survive changing climate conditions, retreats and advances of forest and open country, the uplifting of mountains and flooding of lowlands. Somehow, they not only persist but diversify into an array of wildlife unlike any found elsewhere in the world. People start arriving about 10,000 years ago, bringing new plants and animals with them. Within the last thousand years, Madagascar's largest native animals disappear and its forests recede. Today, forest cover continues to decline, the climate is changing, and the survival of many remaining animal species is in doubt.

I have organised chapters about these events and their causes chronologically, starting deep in the past and ending in the present. The resulting account will doubtless continue to evolve with further research, but it is already far better supported by evidence than the prevailing wisdom about Madagascar. This comes to us from early twentieth-century French colonists, who wrote simply of a timeless, forested paradise invaded and destroyed by people.

Although grounded in up-to-date scientific evidence, I came to realise that this book is also a story. Did the sloth lemur really sing? That, we'll never know . . . My enchantment with Madagascar is lifelong and personal as well as professional, and field experiences and times spent with other researchers are woven into the following pages; the interpretation of evidence is my own and others may interpret it differently; and, most important, knowledge itself comes in many forms and mine is only one of them. I delve into these reflections in the first chapter, and in the last I explore what lessons they may hold for the future.

Around 251 mya:
Ferns and club mosses
replace ancient forests

Around 200 mya:
Primitive reptiles
and pre-mammals

Around 167 mya:
Dinosaurs and
primitive mammals

Around 251 mya:
Embedded in Gondwana;
asteroid hits Earth

200 mya:
Volcanic activity and
global climate warming

Around 7 mya:
Ancestors of endemic
grasses start arriving

Around 2 mya:
Glacial episodes begin
in temperate regions, with
alternating cool and warm
periods in the tropics

Around 7 mya:
Present location
reached, with
modern relief and
climates in place

Around 23 mya:
Sea currents
change direction
and now flow
toward Africa

THOUSANDS OF YEARS AGO (ya)

**Around
10,000 ya:**
Cut-marked
bones from the
southwest signal
human presence

**Around
4,000 ya:**
Traces of a
foraging camp in
the far north

**Around
2,800 ya:**
Traces of a
foraging camp in
the southwest

CENTURIES (CE)

4th–6th centuries CE:
Traces of a fishing
camp, then village,
in the southwest

7th century CE:
Large-bodied
animal species
begin to decline

**8th century CE
onward:**
Villages spread
along coasts

**11th–14th
centuries CE:**
Villages multiply
and spread into the
central highlands;
first major town
and trading port

Around 72 mya:
Amphibians, dinosaurs and many other reptiles, primitive birds and a few mammals, almost none with living descendants

MILLIONS OF YEARS AGO (mya)

Around 125 mya:
Splits with India and Australia from Africa, far south of present location

Around 88 mya:
Becomes an island

65–23 mya:
Ancestors of almost all living animals and many plants arrive

65–23 mya:
Sea currents flow toward west coast from Africa

65–30 mya:
Northward drift through arid belt

66 mya:
Asteroid hits Earth

15th century CE:
Chiefdoms and kingdoms begin to develop

16th century CE:
Large-bodied animal species mostly extinct

18th–19th centuries CE:
Rise and expansion of the Merina Kingdom

The year 1896:
Invasion and declaration as French colony

The year 1960:
Independence regained

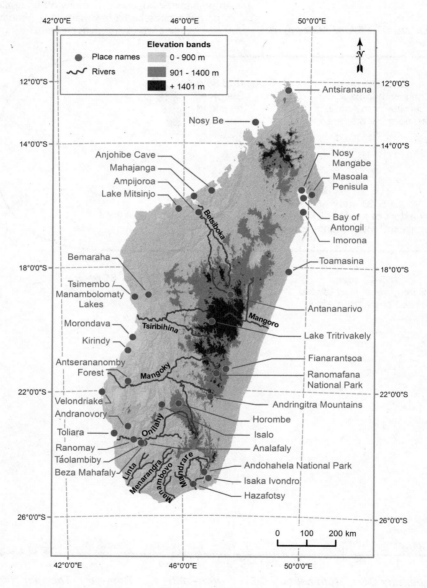

Modern place names mentioned in the text, and major regional cities
(map by Herivololona Mbola Rakotondratsimba)

CHAPTER 1

Living Madagascar

A black and white lemur sings high in the trees of a rain-forest in the east while a tiny brown chameleon falls on its side and freezes on the ground far below, perfectly camouflaged as a dead leaf; a bird in glorious plumage glides between the tips of branches reaching for the bluest of skies above a spiny forest in the south; a baobab tree stands sentinel in a western deciduous forest, its massive trunk a personal water reservoir. This is Madagascar, crucible for the evolution of a grand diversity of life, for plants and animals found nowhere else on Earth.

When I began working there fifty years ago, few people outside the country even knew where it was. Responding to an anxious phone call about me made by my mother in 1970, a Foreign Office official in London inquired whether Madagascar was located in Europe. Things are very different today. The island is now recognised as one of the hottest of biodiversity hotspots in the world, along with regions like Amazonia and the rainforests of Central Africa, and as a poster child for climate change. Through films, television documentaries, newspaper articles and the pages of the *National Geographic*, Madagascar lives in the public imagination as, at once, a treasure of nature and an environmental disaster. Captivating photographs of unique plants and wildlife sit alongside dire ones of burning forests, eroding soils, and headlines announcing the imminent extinction of rare species. Although the island's location is widely known now, what is happening there is less clear.

The delight of almost everyone I encounter who has actually been to Madagascar is palpable. Their pleasure and excitement always puzzle me a little. They are mostly a self-selected group of people with an interest in natural history, and the island harbours many unique plants and animals, to be sure. But it is far from being a fully-fledged tourist destination. Accommodations are often quite basic, and travel is slow, exhausting and fraught. Scattered through this book are sights and reveries during hours scrunched up in an assortment of four-wheel-drive vehicles and, indeed, it sometimes feels as if I have spent more time getting between places than in them. Madagascar lacks great herds of zebra and wildebeest, prowling lions and leopards, or giraffes and elephants to marvel at. Its animals come mostly in small packages, and the sense of sheer good fortune and privilege at catching a glimpse of them doubtless contributes to their magic. But the enchantment goes far beyond that. A closer look at the lemur, chameleon, bird and baobab with which I began helps explain why.

The singing black and white lemur (*Indri indri*) is called *babakoto*, literally 'father of Koto' in the Malagasy language. Some individuals are actually black all over, although those in my mind as I write look like 'a primate version of a giant panda'. They are the largest living lemurs, with exceptionally long hindlimbs that power leaps between tree trunks up to 10 metres apart. A *babakoto* has no tail, just a stubby stump. Lemur tails come in many forms – whip-like, foxy, striped – and it is hard not to feel sorry for *babakoto*, the only species without one. There is no good explanation for the absence.

Babakoto live in family groups – a female, male and their young – in northerly areas of the eastern rainforest. The female lords it over the male, as in many other lemur species. A couple will often stay together for many years while successive offspring grow up and eventually move away in search of mates of their own. *Babakoto* spend much of the day munching leaves and fruit, moving from one food tree to another, or resting. And they sing. A communal roar announces the song, most often in the morning. When the real song begins, the adult couple are the lead singers. The first bars are long notes. After them come wailing stanzas, beginning high and tumbling through several notes to lower wails. The couple sing in duet, although everyone except the youngest animals joins in, and a song may last for several minutes. The sound carries for kilometres across the treetops, and neighbouring groups break into songs of their own in response. Echoes fill the forest with a ghostly lament.

Why do *babakoto* sing? The song is almost certainly a territorial proclamation: 'we are here, stay away'. But it seems to be more than that. The adult female's repertoire includes more notes than the male's, and she moves from one note to the next more quickly. To listeners in other groups, the song in its entirety may communicate who is present – how many individuals, how many males and females. Perhaps it is a way for neighbours to become better acquainted, because direct encounters are rare. Whatever the song's function for *babakoto*, it is surely spellbinding for people who hear it.

Far down below on the forest floor, the brown leaf chameleon (*Brookesia superciliaris*) draws no attention whatsoever to its presence. *Ramilaheloka*, as they are called in some places, are found in many forests in the east, although actually seeing one takes a modicum of luck and a keen eye for small reptiles or, if you are me, being in the company of a good spotter. *They* have less trouble seeing *us*, one suspects: like all chameleons, *ramilaheloka* have the uncanny ability to swivel their eyes backwards and forwards in opposite directions at the same time.

Ramilaheloka are tiny. With bodies no more than 5 centimetres long, they would fit on your thumb. A newly discovered relative, *B. nana*, is even smaller – it would fit on your thumb print! With a vertically flattened body, a *ramilaheloka* looks as if someone had stepped on it sideways. Lying curled up on one side with legs tucked in, it bears a remarkable resemblance to a dead, brown leaf. That is just as well. The forest floor where it spends much of the day catching insects with a long, sticky tongue is a dangerous place for a very small chameleon. A horn sticking up above each eye and spines thrust out from its body may intimidate potential predators close to its own size, but they are no impediment for a hungry bird or snake.

Chameleons are famed for their ability to turn from one bright colour to another. *Ramilaheloka* stay mostly brown or beige. Best not to be noticed. Madagascar is home to many chameleon species, all of them bewitching and many of them larger and easier to see and watch than *ramiloheloka* and its close relatives. The particular attraction of this species, for me anyway, is not just the wonder of an animal passing itself off as a leaf but the mystery still attached to this elusive little creature and the joyful excitement that comes with finding one in the forest.

The crested coua (*Coua cristata*) is called *tivoka* in Malagasy in the south, where I have seen it most often. This vagabond of a bird is quite big, 40–44 centimetres long, and lives in forests of many descriptions up and down the island. It is the most widespread and common of the nine coua species found in Madagascar – and nowhere else in the world. There are no accolades for spotting *tivoka*, and a sighting will not set a serious ornithologist's heart thudding. But they are creatures of great beauty and grace, decked out in a painter's palette of colours. Here is how a distinguished ornithologist describes them: 'pearl gray crested head, neck, chin, and throat. Violet skin around eyes, pearly sky-blue behind eye and surrounded by black line . . . Upperparts green-gray . . . underparts white except for orangey fawn lower chest and maroon upper chest . . . Green-gray wings, midnight blue tail with purple-violet metallic sheen, broadly tipped with white on external tail feathers'. Imagine all that, sailing through the air!

Tivoka sunbathe, and the silhouette of a large bird with ruffled feathers and drooping wings atop a spiny tree limb is a common sight on early mornings in the south. Often, it is alone, and I have never seen more than two together. A *tivoka* is a shimmering singularity, not a flock. It feeds on a cornucopia of insects, and also berries, seeds, snails, and even poor unsuspecting chameleons. The loud, clear calls of *tivoka* belong with late afternoons watching lemurs. '*Coy coy coy coy . . .*' calls one, each syllable a little quieter than the last, and '*coy coy coy coy . . .*' reply others from different parts of the forest. An evening lullaby for me, what do these calls mean to the birds themselves? Like a bird's version of the *babakoto* song, the call may lay claim to a patch of forest or advertise the caller's availability – but that is surmise.

Referring to a baobab as a 'small package' is a stretch, I agree. These iconic denizens of the forest can be giants, soaring 20–25 metres high, with a trunk reputed to hold up to 120,000 litres of water. Australia hosts a single baobab species, and so does Africa. Madagascar far outdoes these continents. It is home to six species of its own, scattered through forests the length of the island's western flank, in addition to the African species that made its way over at some point in the past. Baobab trees (*Adansonia spp.*) are called by several names in the Malagasy language – *bozy, fony, renala, ringy* . . . Here, I call them all *za*, the name by which they are known in the southwest.

Za are designed like no other tree. Short, fat, gnarly branches sprout in an improbable topknot on a smooth, silvery trunk of vast girth. The leaves fluttering on branches during the wet season are undistinguished. Great white, trumpet-like flowers up to 29 centimetres long, tinted pink, red or orange, are this tree's real adornment. With sufficient patience, one also realises that they are hives of industry. Flowers open at dusk, taking as little as a minute or as long as 15 minutes to do so, and only for a single night are they fertile. Glimpses of an array of visitors are the reward for a night-long vigil beside a *za* in flower. Bats (probably *Eidolon dupreanum*) and lemurs (*Phaner pallescens, Cheirogaleus medius*) visit flowers to feast on their copious nectar, carrying pollen from one flower to another while leaving each flower intact. Non-pollinating visitors arrive too, including sunbirds (*Nectarinia spp.*) and a variety of hawkmoths. Sunbirds plunge into the flower and sup nectar through curved beaks while hawkmoths hover at its mouth, using a long proboscis to siphon nectar from deep within. A *za* in bloom is a busy place indeed.

These particular animals and plants offer glimpses of the sometimes secretive, always marvellous diversity of life in Madagascar, and they also happen to delight me. In other words, a mix of scientific interest and entirely unscientific enchantment drove my choices. I am not alone in my fascination. Except for the roustabout *tivoka* – generally seen as fair game especially by small boys with catapults – each is special too in the eyes of Malagasy people living in their midst. But the stories they tell about *babakoto*, *ramilaheloka* and *za* are quite different from my own.

Madagascar is rich in folklore linking people to the wildlife around them – as bringers of bad luck or death, sources of occult power, or fellow creatures to be protected and left in peace. Some stories are about the past, and the origins of Malagasy people. The *babakoto* is a central figure in many stories told by people in the eastern forests. One holds that human ancestors and *babakoto* were brothers long ago, but then one of the brothers decided to come to the ground, live outside the forest and cultivate the land. In another, *babakoto* occupy a special place in human hearts because long ago they helped a child lost in the forest.

Chameleons figure frequently in Malagasy proverbs: '*Hataovy dian-tana ny fiainana, hatreo ny eo aloha ary todiho ny any afara*', admonishes one – '*go about life the way a chameleon walks: face what is in front and look back at what is behind*'. A symbol of wisdom and intelligence for some people, they are emblems of bad character for others, and plain scary for many. *Ramilaheloka* are certainly viewed with suspicion – their name means 'troublemaker' in Malagasy. As for *za*, the Creator made a mistake and planted them upside down, and what seem to be branches are actually roots. Gardening mishaps aside, they are revered everywhere and a lone *za* will be left standing when the rest of a forest is cut down. They are trees of special mystery and magic, and being a source of water adds to this aura. A small hollow carved in the trunk quickly fills with fluid, I learned from a friend as we walked through the forest around the small southern village of Hazofotsy together one day and stopped at a *za* to quench our thirst.

*Tohombinta beside a baobab with a hollow carved
in its trunk, 1971 (photograph by author)*

Long days watching lemurs often ended sitting around with
friends on the mats in my hut in Hazofotsy, my night-time
home for months at a time during my PhD research. One
evening, as a bottle of local rum passed from hand to hand,
the conversation turned to radiated tortoises (*Astrochelys radiata*).
These beautiful great creatures, unique to Madagascar, were
common in the forest. I asked why they were so *fady* (taboo)
that people would not even touch them, let alone pick them
up. The answer was that long ago someone was boiling a
tortoise in a pot over a fire and the pot cracked. No one had
touched or eaten them since.

Returning to Hazofotsy a few years later, the topic came up again and I sagely repeated what I had been told. Much laughter greeted my account. 'Oh, we didn't know you very well then and were too embarrassed to say what really happened, which was that the tortoise rose up out of the pot and bit off the penis of the man tending the fire. *That's* why no one touches or eats them'. Today I remain perplexed about the origins of the tortoise *fady*, but very clear about the sensibilities of my Hazofotsy friends and also, perhaps, the pleasure they take in teasing.

Story-telling is a human characteristic and those told by Malagasy people are part of a web of stories that span the globe. Stories are important. They are how we interpret and make sense of the world. They reflect our values and beliefs, and filter what we choose to highlight from the evidence around us. Take our planet's future prospects as an example. The same body of evidence about current global trends fuels stories with very different endings. Their tellers come in many stripes – scientists, leaders of government and business, not-for-profit organisations and, simply, individuals who care. The darkest stories take the fatalistic view that our species is inexorably destroying other life on Earth and will eventually destroy itself. Our biological destiny is self-destruction. Triumphalist stories strike an optimistic note: human ingenuity and advances in technology will find a way through. Not-as-bad-as-you-think stories emphasise that species have always come and gone and that this crank of the wheel, with our hand on it, is giving rise to the evolution of new species as well as extinctions. Most environmentalists and conservationists embrace yet another theme, one of muted hope and determination: good things as well as bad still happen, and if only we try harder, multiply the good and push back on the bad, all will not end in disaster.

With mats and rum, gravity and laughter, the ways people

make sense of the world in a village of an evening in rural Madagascar seem a far cry from those that depend on scientific evidence as sources. But appearances can be deceptive. The late Malagasy anthropologist Elie Rajaonarison once remarked to me, 'You conservationists are like the missionaries who came here in the past. They came with passionate beliefs, and so do you. Now as then, we accept some of those beliefs and reject others. But make no mistake, Alison, it's not that people don't understand the consequences of clearing forest. It's just that if the choice is between providing food for your family today and saving the forest for the future, there really is no choice'. I remember hotly contesting the analogy between me and a missionary. Still, rolled into a single casual comment over supper, Elie laid bare tensions that haunt conservation efforts in many places – tensions between differing understandings of the world, the kinds of knowledge on which those understandings are based, and the practical choices of daily life.

Early in the twentieth century, French colonists formulated a story about Madagascar that ignored these tensions altogether and offered instead a version of events that went like this: 'Originally, forests teeming with wildlife covered all of Madagascar. Then people arrived a few thousand years ago. In a frenzy of mindless destruction, they cut down and burned 90 per cent of the forest and slaughtered all the biggest animals. Today's rolling grasslands are entirely a creation of Malagasy settlers and their descendants'. This story of changeless forests and wildlife, with a disastrous tipping-point driven by Malagasy people, is flawed at best and plain wrong at worst. Yet it persists in modified form as a widely held view in our own times. Perhaps most startling, 90 per cent is still commonly pronounced as the proportion of forest lost – even though the figure was a guess first recorded a hundred years ago. The figure is used time and again in conservation reports, tourist guides, Wikipedia's description of the island for the general public, and scientific publications – with at least 27 articles published between 1988 and 2011 presenting it as a fact. If the proportion has truly not changed during the last century, that is good news indeed for forest conservation!

A moment in a hot stuffy auditorium in 1970 gave me my first inkling as a young student that there might be a problem with the colonists' story. It was the final session of the first international conservation conference ever held in Antananarivo, Madagascar's capital, hosted by the International Union for the Conservation of Nature. How odd . . . what is a representative of the International Labour Organization doing here, I wondered idly amid a sea of people in the Hilton auditorium. He was Nigerian, I recall, but I do not remember his name and he was not listed as a participant in the conference proceedings. With the heat rising and oxygen level sinking, he came to the podium and said something along these lines: 'From what I have heard during these days, I draw two possible conclusions. One is that the Malagasy are psychologically disturbed pyromaniacs, who get up in the morning with a single question on their minds: what shall I burn today? If that is indeed the case, I suggest we hire a Boeing 747 and bring a planeload of psychiatrists from America to treat them. My alternative conclusion is that Malagasy have reasons for setting fires. In this case, our first step should be to find out what those reasons are.' The audience laughed politely, he sat down, and the conference rolled to its conclusion as if he had never spoken.

At the time, I had not given much thought to the prevailing view of Madagascar's environmental history and its awful tale of human destruction. But those off-key remarks at the conference's end set me thinking that it sounded a lot like the myth of paradise lost, a cultural preoccupation of western societies for thousands of years. 'Myths get made unbeknownst to man' wrote the great anthropologist Claude Levi-Strauss, but that turns out to be only partly true in this case. Mythical trappings gave the colonists' story staying power, but deliberate intent played a role in its creation and set it apart from myth.

I grew up listening to stories. My father was quite old when I was born, and he was a wonderful story-teller. He had lived and worked in Chile for many years as a young man, and I loved listening to him talk about his experiences. Long after his death, I learned more about Chile when he was there, its historic dependence on exports and repressive government. This did not undermine the reality of his stories nor lessen my affection for them, but it did make me realise that there were others to be told. The colonists' story about Madagascar is pernicious, in contrast, and doing battle with it has become increasingly important to me over the years.

Research on lemurs was what first brought me to Madagascar. Amid the coronavirus pandemic, the BBC carries a headline story about the scents male ring-tailed lemurs (*Lemur catta*) produce in the effort to seduce females. My small grandson sports leggings imprinted with a big ring-tailed lemur on his rear. A comedy film about a zoo co-stars John Cleese, Jamie Lee Curtis – and ring-tailed lemurs. I do not altogether understand the popular fixation on this particular species. At our research camp in the southwest, *maki* (as they are called in Malagasy) are the lemur equivalent of bandits, lingering near the table after lunch ready to snatch leftovers. The beautiful, large white lemurs that have always been the focus of my research seem vastly superior to me. Although they rarely leave the forest, occasionally a group will bound into camp like so many ballet dancers, sit down for a while to think deep thoughts, and then

*Author taking a break with the late Georges Randrianasolo and his nephew
in the forest at Ampijoroa, 1970 (photograph by Paul Godfrey)*

be on their way again. Their scientific name is *Propithecus verreauxi*.
In the Malagasy language, they are called *sifaka*, a name that
sounds (a bit) like their alarm call. I dare say my preference for
sifaka over *maki* is biased, although the two species certainly have
very different temperaments, perhaps suited to the way each
makes a living in the forest.

Yet *maki* and *sifaka* share a feature that is widespread among
lemurs and very uncommon in other primates, and it was one
of the attractions of Madagascar for me. I was interested in the
evolution of social systems, and lemurs, like many of Madagascar's
animals, break a lot of evolutionary 'rules'. Studying the circumstances under which social rules get broken helps explain why,
for the most part, they hold. One broken rule particularly
intrigued me. Unlike most primates, lemur females of many

species are socially dominant to males while their primate sisters elsewhere are busy subordinating themselves. For my PhD, with considerable hubris in retrospect, I set out to understand why this odd characteristic evolved. On countless occasions, I watched male *sifaka* waiting patiently while a female ate the best, most delicious fruit, flowers or leaves in a tree. Only when she moved away did he take her place and eat whatever remained, and woe betide the male who made a mistake and approached prematurely. Met with a glare or even a cuff of the hand, he retreated snickering with anxiety, tail curled submissively between his legs.

Why did this unusual asymmetry evolve? I wish there were a definitive answer after many years of research, but there is not. Female social dominance probably evolved among many (though not all) lemur species because having offspring is particularly burdensome and ready access to the best food essential, and so natural selection has favoured females who face down males, and males who are wimps. And why is reproduction particularly burdensome among many (though not all) lemur species? The answer grows vaguer, and probably involves idiosyncrasies of both lemur physiology and Madagascar's environment. But perhaps there is no single way of 'solving' the 'problem' of social relations between females and males, and evolution simply found an unusual solution in Madagascar.

The idiosyncrasies of Madagascar's animals and plants offer a wonderful array of rabbit holes down which a person fascinated by the natural world could disappear for a lifetime. I have spent a good chunk of mine doing just that. Extracting myself to contemplate the landscape above-ground, so to speak, usually happened thanks to my late husband Robert Dewar. Bob and I met at Yale in 1972, when I was a newly minted assistant professor and he was finishing a PhD on early agriculture in East Asia.

China was not open to foreign archaeologists at the time, and Bob was looking for other places to work. Madagascar is full of interesting questions, I pointed out – how about there? Somehow, both Madagascar *and* marriage followed a few years later.

Bob's central interest was in how people's activities in the past shaped the landscapes they occupied. Madagascar turned out to be a really interesting place to pursue this, and Malagasy archaeologists welcomed the addition of another trowel-wielding colleague and, over time, friend. Slogging up hillsides to archaeological sites together, co-authoring articles, talking at the kitchen table, Bob vastly expanded what I thought about and, in particular, gave me a lifetime interest in the history of human settlement and what happened before people arrived.

Time out in the north, with Bob and the children, 1991
(photograph by Celeste Peterson)

A further reason for returning to Madagascar was soon added to my own research and the work Bob was doing. I saw forests and wildlife disappearing. Pitching in to do what I could to help conserve the island's natural habitats and wildlife quickly became an urgent priority. Inspired by the president of the School of Agronomy at the University of Antananarivo, in 1974 I teamed up with him and his colleagues, a colleague at Washington University in the US, and the leaders of Bezà Mahafaly, a village community in the southwest. We forged an unlikely partnership. Together, we all agreed, we would find ways to protect the forests and wildlife in the area while at the same time improving local people's livelihoods and well-being. It was grassroots, community-based conservation before the term was invented – and the start of a long and continuing education for me.

For many years, these threads seemed to belong to separate lives – my own research in biological anthropology and Bob's in archaeology, and another far removed from academia in a world of practical action. The separation in my mind reflected a conventional view at the time of what academic life could encompass. The boundary has become more permeable since then and the split in my own mind has disappeared. All the threads of my life are interwoven in this book. It explores Madagascar's long environmental history, the challenges facing the island today and the possible futures ahead, based on the best, most recent evidence available. But it is also my story, the evidence filtered by my experiences and by beliefs and hopes I am sometimes aware of and sometimes surely not.

Tsingy de Bemaraha National Park (Michail Vorobyev / Shutterstock)

CHAPTER 2

The Rock That Moved

Landfall! I've flown to Madagascar more times than I can count, but still feel a lurch of excitement when the coastline appears. From the aeroplane window I watch it come into sight, a great red stain in the sea heralding its approach. The African coast disappeared from view an hour or so earlier, several hundred kilometres to the west, and reaching other major landmasses around the rim of the Indian Ocean would take far, far longer. India lies 3,800 km away, Indonesia 5,600 km and Australia furthest of all, 6,800 km away.

Madagascar is big, with a surface area of approximately 590,000 km^2 – about the size of France, Belgium and Luxemburg put together – making it the world's fourth-largest island, after Greenland, New Guinea and Borneo. Its combination of great size and wide variety of relief and climates gave rise to one of my favourite article titles in a scientific journal: 'Madagascar: heads it's a continent, tails it's an island'. But it wasn't always so. Like every landmass, Madagascar has been on the move and changing for hundreds of millions of years, its own particular history inextricably linked with that of Earth itself.

Movement brought new spatial relations with other lands, new boundaries between land and water, and new climates. Geological upheavals and the slow weathering of rocks reconfigured its landscapes repeatedly during the long journey. Unfolding events helped determine the plants and animals present at a given moment in time, which had the best chance of reaching Madagascar after it became an island, where they were most likely to come from, the conditions they would face, and the likelihood they would survive.

Describing the physical history of Madagascar is a prelude to exploring the history of life there. Where did Madagascar's travels take it, what company did it keep, and what ocean currents, winds, and climates did it experience along the way? What was happening on land, and when and how did Madagascar become a place of dramatically varied relief and eroded landscapes? Some clues to the answers come from studies of Earth's crust, others from the geology of Madagascar itself, and still others from climatology, palaeoclimatology and models simulating ancient events. Together, they bring us to the present from deep in the past.

The terrain is familiar as the aeroplane comes in to land at Antananarivo's airport. It all looks so solid, and I remind myself that this is more a reflection of the brevity of human life than of the changelessness of Madagascar itself. Trudging wearily across the tarmac after the long flight, dark thoughts about mortality quickly take second place to making it through immigration and reconnecting with my luggage, finding the colleagues-become-friends patiently awaiting me, exchanging sleepy gossip on the long drive into town and finally, blessedly, falling into bed. Another field season begins – with a good night's sleep.

*

I belong to the first generation that came of age when the drifting of continents was becoming a widely accepted fact. It is difficult to imagine how perplexing the modern arrangement of continents and the biological life they hold must have been for geologists and natural historians who came earlier – and not much earlier: '. . . if Gondwanaland existed, Madagascar was part of it' wrote one geologist as recently as 1972.

Alfred Wegener's name was well known to me as an under-graduate, for he was the star of the continental drift saga during the first decades of the twentieth century. The line between adventurer and academic was fuzzier then. Wegener was an explorer with broad interests in the natural world. Wegener-the-academic was a geophysicist, meteorologist and astronomer, best known in his lifetime for groundbreaking polar research. Today, however, he is celebrated for insights that ultimately transformed our understanding of the workings of our planet.

Pondering a map of the world, like others before him Wegener was struck by an obvious fact: the east coast of South America and west coast of Africa look like pieces in a giant jigsaw puzzle waiting to be put together. But Wegener was different from others. Slowly and painstakingly, he assembled geological evidence showing that continents had once been joined to one another. The question was how they became separate. Some argued that land bridges formerly connecting them had sunk, while Wegener hypothesised that continents drifted apart. He was right about this, even though his particular ideas about how it happened turned out to be wrong – and attracted considerable derision at the time.

We now know that continents drift because plates of crust encircling the globe glide slowly over an inner layer covering the Earth's core. These plates carry with them continents that rise above the sea in places where the crust is particularly thick. It took advances in technology, accumulating evidence, and long, contentious debate to gain acceptance for what geologists today refer to as the theory of plate tectonics. Wegener and others at that time were missing information about a key component of the system: ocean basins. Hidden deep under water, crust on the ocean floor is the conveyor belt controlling the movement of continents. Ocean crust is thin and dense, very different from the thick, light crust of continents embedded within it. It is constantly created in deep centres found in every ocean. New crust pushes older crust away, widening the ocean floor, but oceans cannot widen forever on a planet of fixed size, and so ocean crust is also constantly cycled back into Earth's interior at what are called subduction zones. Continents drift passively across the Earth's surface (moving 1–10 cm per year), driven by tectonic processes on the ocean floor.

Madagascar played an unremarked, practical role in the development of ideas about all this. By the time the fourth and final edition of his treatise on continental drift appeared in 1929, Wegener had made much progress. Among other things, he presented estimates of relative rates of drift for eight land-masses and calculated changes in the geographical coordinates of two directly – Greenland and Madagascar. By his estimate, Madagascar moved 60–70 metres east per year relative to the Greenwich meridian (0° longitude) between 1890 and 1923. Scaled up in time, the implication was that Madagascar moved 40,000 kilometres east over 20 million years – which would actually mean drifting east across the Indian Ocean, the Pacific Ocean, and the Atlantic Ocean to more or less where Africa lies today! Wegener clearly understood the improbability of this and noted the need for more measurements, while adding cheer-fully that 'the observed longitudinal change of Madagascar is in the right direction to fit drift theory'. Whatever the correct figure might be, Wegener concluded that Madagascar, along with all the Earth's landmasses, had been on the move throughout history.

How much Wegener and his colleagues got right is more remarkable to me than what they got wrong. Tracing ancient movements of Earth's surface is hard even with sophisticated technology, and modern scenarios are buttressed by a range and quality of evidence unimaginable in earlier times. The best way of visualising events may be to watch an animated version, but words paint another kind of picture.

The supercontinent of Gondwana came into existence around 500 million years ago, composed of over half the planet's landmasses with Madagascar embedded in their midst. The array of lands in the immediate vicinity has an improbable ring today – East Africa, the Seychelles, India, Sri Lanka, Australia and Antarctica. The reality was humdrum. Madagascar was a modest wedge of land scrunched between many neighbours. It became even more inconsequential when Gondwana collided with North America, Europe and Siberia to form Pangaea. This *super*-supercontinent, composed of all the world's continents surrounded by a global ocean of water, persisted for well over a hundred million years.

After the breakup of Pangaea, Madagascar continued to ride along incognito inside Gondwana. Big continents have harsh interiors. Far from the moderating effects of water, parching hot summers would have alternated with bitterly cold winters in ancient Madagascar, and it must have been as cold and bleak a place as it would ever be when Gondwana sojourned over the South Pole for a while. Madagascar's colours during that period would have been unremitting shades of grey rock and white snow, a far cry from the palette we know today.

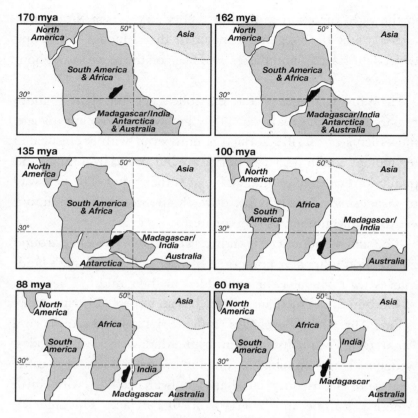

Madagascar's long journey to becoming an island
(redrawn from Yoder & Nowak 2006)

Gondwana was still well south of the Equator when it began breaking apart about 170 million years ago. One fracture opened a seaway between Madagascar and East Africa around 125 million years ago, today called the Mozambique Channel, and Madagascar drifted away east and south. It was not yet an island, though. Australia and Antarctica were still connected to the south, and India's attachment ran the length of the eastern coastline.

*

Although the timing of Antarctica and Australia's split from Madagascar remains uncertain, India was probably the last remnant of Gondwana to drift away. That happened between 80 and 88 million years ago, and after that Madagascar was completely surrounded by water. Alone at last – probably. New studies of the Mozambique Channel's floor hint that a land bridge may have intermittently connected Madagascar to Africa, but the jury is still out about this. As the other continents continued on their separate ways – India northeast toward a collision with Asia, Australia southeast, and Antarctica toward the South Pole – Madagascar embarked on a stately passage to its own present location. Starting out far south of where it is now, it began moving northward in tandem with Africa about 65 million years ago. The ensuing journey was long as well as slow – 1,650 km across 15 degrees of latitude.

Paddling in the Mozambique Channel as waves lap gently up a beach on the southwest coast is a pleasant way to spend an afternoon. The sea and sky vibrate in blue, a sharp white line of breaking waves marks the coral reef a few miles out, pale sand shimmers along a curving beach as far as you can see, and the horizon stretches forever. It is a place to contemplate the consequences of being surrounded by water.

Ocean currents and winds shape these consequences in important ways. They exert a strong influence on climate. They determine how easily creatures able to fly or swim can cross a barrier of water. They set the odds of land-living animals successfully traversing open seas aboard floating rafts of vegetation. They affect the chances that people in sailing boats or canoes will make landfall. They are key to understanding what isolation has really meant for Madagascar.

Unravelling their history, it is easiest to start with the system of winds and currents in the Indian Ocean today and then look back into the past. Warmed by the sun, especially in equatorial regions, Earth's surface heats the air, causing it to rise. Pushed outward as well as upward by newly warming layers beneath, the air drifts away north and south toward the poles. Cooling as it drifts, the air releases moisture in the form of rainfall. Eventually it cools enough to sink again and the dry, descending air creates an arid belt circling the globe at mid-latitude in each hemisphere. Air from these belts flows back toward the Equator across Earth's surface, where the cycle starts again. This circulation system is the origin of the year-round trade winds that blow in northern and southern stretches of the Indian Ocean.

Mirroring circulation in the air, gigantic eddies, known as gyres, churn away in the waters below. The northern gyre moves clockwise, flattened and weakened by India jutting into its path. The southern gyre spins counterclockwise. Big and powerful, with no land all the way to Australia to blunt its force, it is like a cog between the northern gyre and the southern current that races around Antarctica. The two gyres converge on the far eastern side of the Indian Ocean and form the mighty Equatorial Surface Current, which streams westward across the ocean toward Africa – and Madagascar.

The Indian monsoon system interacts with and complicates these grand gyrations. Seasonal differences in the heat of the sun and its warming effect on lands north of the ocean cause winds and surface currents in these reaches to reverse direction. Between December and April, a steady wind out of the northeast blows from the Arabian Sea toward the African coast. In April, the system reverses and the winds blow out of the southwest. The northern tip of Madagascar today is licked by the northern gyre and monsoon system. To the south flows the fierce current circling Antarctica. The east coast is buffeted by the Equatorial Surface Current surging across the Indian Ocean.

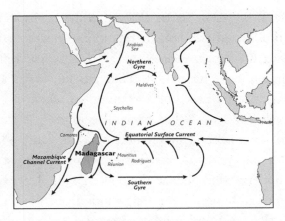

Ocean currents around Madagascar today
(adapted from Boivin et al. 2013)

Along the west coast, dangerous currents race southward down the Mozambique Channel. Linear distances alone do not capture the extent of Madagascar's isolation for those without aeroplanes or motor-powered boats to get there.

Sixty-five million years ago, the position of Madagascar and continents around the rim of the Indian Ocean generated very different dynamics of water and wind. Far south of its present location, the east coast of Madagascar was not yet exposed to the direct force of the Equatorial Surface Current. The current hit the longer African coastline and sometimes bounced back east, acting as a sort of watery umbilical cord connecting Madagascar to Africa. That changed roughly 23 million years ago. Madagascar had drifted far enough north by then to be in direct line of the current. The current now hit Madagascar first, before swirling around the island toward Africa and dissipating north up the African coastline or south through the Mozambique Channel. The umbilical cord was cut, with profound consequences for the history of the island's wildlife.

*

The climate of the young island altered dramatically as it drifted. Shifting latitude and position in relation to nearby continents, together with the emergence of varied relief across the island drove much of what happened. Madagascar was not only far south of where it lies now but also quite flat when it started out 65 million years ago on the last stage of its journey to the present. It seems to have been a green and pleasant land. Plants and animals from that era discovered in the northwest suggest a climate similar to modern conditions in the region, with plenty of rainfall in most years, although prevailing winds from the west would have made the west and south wetter and the east drier than today when conditions are dominated by easterly winds.

The northward drift soon brought Madagascar to the arid belt. The belt was wide, wide enough to swallow even a big island, and harsh, dry conditions enveloped Madagascar for tens of millions of years. The well-watered place of the journey's outset vanished. The island's northern tip emerged first from the arid belt into balmier tropical climes and, by about 30 million years ago, only the south remained within the belt's grip. And there the southern tip still lies today.

Sojourning in the arid belt on the long drift toward the Equator is only part of the story. The island's relief also played a crucial role in determining its climate, and the smorgasbord of rocks on view today holds clues to the dramatic upheavals and slow processes that produced the varied landscapes and climates of the present. The land was lower as well as flatter 65 million years ago, and the whole island may have been uplifted 1–2 kilometres by movements of the Earth's crust in just the last 15 million years. Rugged mountains running the length of the island in the east record its time in Gondwana's interior. They expose the eroded roots of one of the seams along which Gondwana was sewn together, and the straight east coast matches the equally straight west coast of India from which it split. Many rocks in the east are very old indeed, their crystals forged under intense heat and pressure in Earth's crust when the supercontinent first assembled. Some are younger, formed from lava and ashes spewed out by volcanoes after Madagascar became an island.

Little of Gondwana is visible in the west, in contrast. The sea surged in during the split from Africa, submerged the land, and buried ancient rocks under deep deposits of sediment. Over time these sediments became limestone, forged from skeletal fragments of marine creatures, and sandstone, composed of sand-sized grains weathered out of older rock. When sea levels fell, the west became dry land again. Some of the limestone is less than 60 million years old. This land is 'young'.

Madagascar's relief became dramatically asymmetrical as a consequence of these events. An east–west slice through the middle would show mountains composed of ancient rock soaring to 2,000 metres near the Indian Ocean in the east and dropping gently across the younger, rolling hills of the central highlands, with flatlands stretching to the Mozambique Channel in the west. Great rivers with large deltas at their mouths drain two-thirds of the island's land surface into the Channel. Heavy downpours in the rainy season briefly turn them into raging torrents, but for most of the year they meander across the western landscape in a leisurely way. Rivers in the east are hares compared to the tortoises of the west. Reaching the eastern edge of the central highlands, the road winds down beside rushing streams and waterfalls until the land levels out. It is a precipitous plunge, with no space for wide deltas on the narrow coastal plain. The descent from the central highlands to the south is gentler, and a single sweep of the eye simultaneously takes in the old and the new. The craggy ramparts of the Andringitra mountains loom grandly to the east, approaching the age of Earth itself, while younger, rolling lands stretch to the horizon in the west and south.

Picture Madagascar around 8 million years ago – almost free of the arid belt, its relief increasingly varied, and neighbouring landmasses more or less where they sit now. The foundations of today's climates were in place at last, and the island would have been far more recognisable to us than in earlier epochs. Year-round, the Equatorial Surface Current carried tropical warmth and moisture to the east, and southeast trade winds arrived laden with water, while the monsoon system brought seasonally warm, moist air to the north and central highlands. Relief did the rest. Forced up by mountains in the east, the warm air cooled abruptly and lost its ability to hold moisture. The result was heavy rain-fall, and the eastern lowlands became a dramatically wet, warm region. The winds blew on westward, losing what remained of their moisture to rainfall as they went, with little left at all for most of the year by the time they reached the west coast. A north–south gradient skewed the pattern further, with humid conditions in the northwest tailing off into near-desert aridity in the southwest and far south. Grand and complicated relationships playing out between the positions of landmasses, winds, currents and relief produced 'local results'. Then as now, these added up to an island of continent-like climatic diversity.

But the saga does not end there. Starting about 2 million years ago, global fluctuations in temperature led to periods of glacia-tion at high latitudes, and to alternating warm and cool phases in the tropics and subtropics. The last glacial episode ended 12,000 years ago, although subsequent climate shifts brought drier conditions to northeast Africa and large areas of the Sahel. The environmental consequences were not trivial, for this was when the grasslands of northern Africa turned into the Sahara Desert. These global events echoed across Madagascar too, and I return to them in Chapter 8.

The difference between climate and weather is a matter of scale. Climate is the sum of weather playing out over decades, centuries and millions of years. The fine-grained time scale of weather – days, months and years – makes it impossible to reconstruct in the distant past. But weather is important. It is what animals, plants and people contend with day to day and year to year, and coping with whatever the weather throws at you can be the difference between life and death. Regional weather in Madagascar today, very different in most respects, shares a common feature: patterns of rainfall between or within years are unusually unpredictable. It is not yet clear when this curious feature was established, though it is likely as old as the general conditions supporting modern climates. Pinning it down in the present and exploring its evolutionary implications preoc-cupied Bob and me for many years.

I know a lot about unpredictable weather. Take two excursions to the village of Analafaly in the southwest, for example. In January 2012, I set out from our camp on the edge of the Bezà Mahafaly Special Reserve with Andry Randrianandrasna, then-head of the reserve – and Chief Optimist – to visit Analafaly. Our camp lay in forest west of the river, and the village a few kilometres east of it. Though the rainy season was underway, Andry assured me the river would be ankle-deep. No problem. *No problem?* The current was fast and the water well above my knees as I carefully edged along. Midway across, Andry looked back, grinned, and shouted 'Remember to watch out for the holes!' My edging slowed further. People dig deep holes in the river's sandy bed in the dry season when it is empty, and their cattle drink the water that collects at the bottom. People drank that water too, before village wells were dug.

Emerging from the river in triumph, all holes avoided, we followed the path to Analafaly across fields on the river floodplain.

My *lambahoany* flapped wet against my legs, but the sunshine and hot wind quickly took care of that. The fields were festooned with green shoots of maize, carefully inter-planted with sweet potato. Men and women busy weeding and planting paused to call out greetings and exchange gossip and jokes as we walked. The harvest would be abundant this year.

January 2014, and another visit to Analafaly. This time, the river was bone dry. We padded across its sandy bed in silence. Desiccated maize stalks rattled in the hot wind. One man hoed a field where the maize had not yet quite died. He looked at us but said nothing. Otherwise, there was no one there. There was nothing to do. The greetings were serious and anxious when we reached Analafaly, and no jokes were bandied about. The village elders had invited us to visit in order to ask for help in securing food aid for their children. The rains had failed, and the harvest would be meagre. Farming is tough when you do not know if the rains will fail from one year to the next, and the impact of global changes in climate is likely to make these failures more frequent.

Author and friends emerging from the Sakamena River in flood, 2012
(photograph by Sibien Mahery)

Intermittent droughts are not uncommon in the world, particularly in dry regions, but it was Bob's hunch that Madagascar was a hotspot of unpredictable droughts – and not just a source of weather stories told by his wife. In 1996, he and a colleague assembled decades of rainfall data from almost 1,500 meteorological stations around the tropics, looking for evidence of this. Their analysis examined the likelihood of a major shortfall of rain in any one year in a given decade, and grouped the resulting probabilities as low, average, or above average. Eastern Brazil, the far north of Australia, certain Pacific islands, and the coasts of Africa all turned out to be hotspots of high and unpredictable variability.

Only two meteorological stations in Madagascar had sufficiently long and accurate daily records to be included in the study. Often, it seemed, the paper or pens ran out or a cyclone so disrupted life that no rainfall was recorded even as rain drenched the area. Of the two stations, only the one in the southwest provided evidence of high year-to-year variability. This was disconcerting. Perhaps the whole idea was a red herring.

Deciding to plug on and investigate further, Bob compiled the best rainfall data possible from an additional 13 stations in Madagascar. The new sample confirmed that variability between years is not a general characteristic of the island. He found it in the north and southwest, but data from stations in the east all fell within the global mid-range. This led us to think about a different kind of unpredictability and pose a new question: at stations where the total volume of rainfall varies little between years, how predictable from year to year are the particular months when rain falls?

We compared the data from Madagascar with a sample from almost 600 stations with similar average rainfall in Africa.

The result was clear. The monthly pattern of rainfall in eastern Madagascar was strikingly unpredictable from year to year compared to Africa, even though a similar total volume of rain fell each year. Globally, unpredictable variability tends to be high in regions near the tropics that have low average annual rainfall, though why there are clusters of hotspots and why Madagascar is one of them remains a mystery.

Unpredictability increases the uncertainty of life in Madagascar, but hurricanes, or cyclones, are its devastating exclamation marks. A map of the paths taken by cyclones in the Indian Ocean in recent years appears, indeed, to bury the island. November to April is cyclone season. Between 1980 and 2007, 12 or 13 cyclones formed in the ocean east of Madagascar on average each year. Averages can be deceptive though, and the frequency and intensity of cyclones likely increased during this period.

Forty-eight of the 92 cyclones arising in the waters around Madagascar between 1980 and 2007 made landfall. Mostly, they barrel ashore along the northeast coast, but from time to time they unleash assaults around the entire coastline and their fate varies once they arrive. Some weaken quickly and fade away. Others linger for days as tropical storms, or make their way across the island, bringing high winds or rain, or both. For the eastern forests in particular, cyclones are a source of destruction – and diversity. Flattening vegetation in their path, they reset the ecological stage, opening up opportunities for new arrivals to take root and flourish. It is uncertain when frequent cyclones became a feature of life in Madagascar, but probably, like unpredictable rainfall, it emerged with the modern climate system around 8 million years ago.

*

Madagascar continues to reverberate with the activities of the plate on which it sits. As many as a hundred small earthquakes each year, active volcanoes, and hot springs all serve as reminders that it still has a geological heartbeat. Arriving in Antananarivo late one evening in August 2016, I fell gratefully into bed. That night, an earthquake shook the city and my hotel building trembled. It did not do much damage, but all the conversation next morning was about things that fell down or broke and the fearful feeling of being in the clutches of an angry Earth. I slept through the whole episode and could not decide if I was pleased or disappointed.

Weathering and erosion are less obviously dramatic than the Earth's crust in motion, volcanoes erupting and seas swirling, but they have had a deep influence on what we see today. Weathering refers to the breakdown of rocks in a particular place, erosion to movement of the products of weathering between places. Both involve the actions of water, wind and temperature. Among the many beautiful sights in Madagascar, limestone landscapes in the west are among the most stunning. They are products of weathering and erosion. Worn down and hollowed out by seeping water over millions of years, the limestone formed canyons, crevices, caves and soaring, needle-like pillars called *tsingy*. Today, the largest of these formations are protected as national parks, and the *tsingy* of Bemaraha have been named a World Heritage Site. A steady flow of tourists come to visit and admire these places.

Weathering and erosion take many forms, however, and they have an entirely different look in the central highlands. There, the marks of recent erosion are heart-shaped gullies, brick-red from freshly exposed soil and streaked orange and white by the bedrock beneath. The red earth is called lateritic soil, its vivid colour a marker of high iron oxide content. Lateritic soil is formed by the chemical weathering of ancient rock, and is common in tropical areas. The gullies' crumbly bedrock, also formed by chemical weathering, is a thoroughly decomposed and porous rock called saprolite.

The spectacle of gullies (called *lavaka* in Malagasy) is inescapable across the undulating hills of the central highlands. They can be huge, 75 m wide, 300 m long and 70 m deep, and in some places the total area of *lavaka* exceeds that of uneroded hillslopes. The land is 'terribly gullied'. They are a daunting sight, and it is easy to see how *lavaka* have stoked metaphors of Madagascar as a wounded, bleeding body. Environmental reports, television documentaries and travel guides take them as evidence of human destructiveness, moreover, eliciting not admiration but calls for changes in the way land is used. But the idea that deforestation, overgrazing and burning are the sole or even main 'culprits' rests on problematic assumptions and scant or outdated evidence.

In recent decades, American Neil Wells, Malagasy Benjamin Andriamahaja and Irish Ronadh Cox – a truly international trio of geologists – have convincingly shown that natural processes are also at work. My path has crossed with each of theirs in the serendipitous way such intersections often happen in Madagascar. Neil collaborated with Bob many years ago, and that brought us together from time to time. Benjamin went on to become a mainstay of Ranomafana National Park and troubleshooter-in-chief for overseas researchers setting up projects in Madagascar, and we became good friends when his daughter studied at the Yale School of the Environment. Ronadh, now immersed in studying the enigmatic habits of huge boulders that inexplicably roll up beaches on the west coast of Ireland, generously came to my rescue as I struggled to master the complexities of Madagascar's geology for this book. The work they and their collaborators have done on *lavaka* is in the spotlight here, although for me our lives feel intertwined in other ways too.

Lavaka have life cycles. They mostly start in mid-slope on hillsides curving gently outwards and covered by thick red soil over crumbly bedrock. A crack in the protective soil exposes the bedrock to erosion, and large, deep gullies can develop very quickly. This usually happens during the wet season when rainfall is heavy, the earth saturated, and groundwater flowing freely. Soils and eroded bedrock collapse into a central hollow with vertical walls, and mud washes out into streams. Eventually the *lavaka* stabilises, as earth falling from its walls fills in the hollow and the walls themselves become more gently sloping. Lichen and grasses gain a foothold, shrubs and trees follow, and over time vegetation conceals the raw scar completely.

Is human activity the only or most important trigger in their formation? To be sure, cart tracks, footpaths and cattle trails promote *lavaka* by denuding and compacting the ground, thereby concentrating and hastening runoff. Hill fields, leakage and overflow on the sides of canals and ditches and, in centuries past, defensive trenches around fortified hilltops add to the list of human activities that certainly contribute to erosion today. Yet the evidence gathered by Neil and Benjamin painted a nuanced picture. Examining 110 *lavaka*, they concluded that only a quarter could be directly linked to human activity; another quarter, they judged, were the result of natural causes; and in half the cases the cause of erosion could not be determined. The lack of grounds for assigning most or all the blame to Malagasy farmers and pastoralists was clear, but equally clear was the need for more research.

Ronadh plunged in. If human activities were primarily

responsible for *lavaka*, she reasoned first, they should be concentrated where those activities are most intense. This turns out
not to be the case. With a sample of 61,000 *lavaka*, she showed
that clusters tend to occur in places where small earth tremors
are common. Shaking the ground gently and repeatedly is
evidently a good way of setting the process of *lavaka* formation
in motion. Although this made clear that natural processes were
often involved, it did not establish the actual age of *lavaka* or
whether any predated the arrival of people. The precision of
measurements declines the further back in time one goes, and
determining the existence of ancient *lavaka* posed large and
tricky technical challenges.

Isotope analysis came to the rescue. The atoms of a chemical
element have variable numbers of neutrons, and the different
forms are called isotopes. Some isotopes are stable; others are
unstable, and decay to a stable form at a predictable rate. One
such is ^{10}Be. An unstable isotope of beryllium formed by cosmic
rays, it decays very slowly and this allows measurements over
long periods of time. It is found in sandy sediments, and different
sediment types have differing concentrations of ^{10}Be, called
signatures. The ^{10}Be signature is high near the surface of the
ground, which is heavily exposed to cosmic rays, and low in
deeply buried sediments.

Ronadh and her colleagues found that while the ^{10}Be signature of sediments on the surface of hillsides was high, that of
sediments buried in many *lavaka* was low, signalling that they
had long been buried out of reach of cosmic rays. This new
evidence strongly suggested that the hill slopes of central
Madagascar were pockmarked with *lavaka* far in the past, with
no sign of a major increase after the arrival of people.
Questions linger about the impact of human activities, however,
and ongoing work should shed new light on this. But it is clear

that *lavaka* became part of the landscape several million years ago, when Madagascar's modern configuration of geology, climate and earthquake activity fell into place. *Lavaka* belong to Madagascar's long geological history as well as the present. The particular combination of ancient bedrock, soil, relief, climate and weathering in the central highlands makes them common there, and *lavaka* make a disproportionately large contribution to the sediment that turns Madagascar's rivers red.

The work to refine estimates and understand the processes that drive erosion continues. Erosion rates in Madagascar are commonly said to be unusually high compared to other regions of the world. But recent estimates show this is not so, and the *lowness* of rates is what seems surprising. The dramatic red stain in the ocean off the northwest coast conjures awful losses from *lavaka* far inland, yet the sediment flushed out to sea by rivers after storms is nothing out of the ordinary, even though it is visible from space. The rate at which sediment is created in the highlands is no different from the Appalachian Mountains in the US, and sediment levels in Madagascar's western rivers – supposed exemplars of people-driven erosion – turn out to be in the middle of the pack of rivers worldwide. Even the highest reliable estimates fall far short of the global midpoint.

Madagascar's erosion performance is distinctly below average when actually measured, but *lavaka* and the red-stained river mouths of the west surely win a prize for visual drama. It is perhaps for this reason above all that they have come to be a potent symbol of Madagascar's environmental problems in our own times.

The village of Ranomay sits on the banks of the Onilahy River in the southwest, at the bottom of a steep hill beside Route

Nationale 10. Bob and I stopped there with the Bezà Mahafaly team on our way back from several weeks in the field together in August 2012, because we knew the villagers produced salt. Patches of salty soil are widespread in the southwest, a reminder that the parched and arid terrain was mostly formed under the sea. Our mission that day was to learn more about methods of salt production. But it was paddling in hot water that made the day truly memorable.

After a long tour of the salt works, the village president kindly invited us to visit Ranomay's hot springs. This surprised me: the hot springs of the central highlands are famous, but I did not know that here too, in this utterly different landscape, there would be evidence that the landmass of Madagascar is 'alive and cooking'. I should have realised, of course: the village's name, Ranomay, means hot water in the Malagasy dialect of the southwest.

A low dam collected the water bubbling out of the ground into a shallow pool. Beside it, what at first glance appeared to be a chapel turned out to be an ornate but crumbling stone bathhouse, dating from colonial times. The people of Ranomay were repairing it in the hope of attracting tourists. It was an altogether improbable sight in this remote corner of the island. Steam rose from the crystal-clear water, and it seemed sacrilegious to wade in with dirty, dusty feet. But we did. Hot feet in hot water on a hot day felt good, and conjured a living, moving landmass in a very direct way. A small piece of Gondwana became one of the world's biggest islands; flat lands became mountains; the spatial relations between Madagascar and other lands changed, and shifting winds and waters exacerbated or diminished their effective distance from one another and the difficulty of reaching Madagascar from them. It is hard to fit the landscapes of the present into the same framework as those that existed hundreds of millions of years ago. Yet they belong together, and Ranomay somehow made that clear.

The mighty Beelzebufo ampinga *alongside a modern frog*
(drawing by Luci Betti-Nash)

CHAPTER 3

Life in Deep Time

Thursday nights in Antananarivo used to be punctuated with the rumble of wooden wheels on cobblestones, the steady clump of hooves and occasional moo, bursts of laughter, murmurs of conversation, and thuds as heavy sacks hit the ground. They were the sounds of oxen pulling carts laden with produce and goods, and of people converging on the city from the countryside ahead of the huge, sprawling market that sprang to life in the city centre every Friday. In order to relieve the weekly chaos and congestion, the market was moved to the edge of town some years ago and Friday is like any other day now.

I loved the sights and smells of the old market, and especially the section with minerals for sale. Small stalls displayed a breathtaking assortment, from rough-hewn, glittering lumps of rock to finely worked gemstones, but my eye was always caught by thin, roundish slices of stone polished to a high gloss. Concentric circles of brown, burgundy, gold, silver and cream flowed within the narrow grey rims of these beautiful objects. They were

49

petrified wood, I was told, cross-sections of ancient tree trunks. I marvelled at these fossils. Writing about Madagascar's distant history decades later, my feelings are more complicated. Beautiful as they might be, those stones were pieces of Madagascar's past sliced up for sale to tourists and people like me.

What kinds of plants grew in ancient Madagascar, and what kinds of animals roamed the land? Clues are hard to find, but palaeoclimatologists, palaeobotanists and palaeontologists are great sleuths and this chapter draws on their work. Above all, the clues tell us that ancient Madagascar was filled with strange, vanished forms of life, utterly different from those of the island we know today.

*

The earliest evidence of plants comes from fossil tree trunks around 300 million years old, slices of which I admired so much in the city market. These trees belonged to an ancient plant group that evolved long before flowering plants, a group whose surviving descendants include ginkgoes and pine trees. Coal deposits in the southwest are distant echoes of this period. They signal forests – it takes a lot of dead plants to make a coal deposit – but most of the tree species were quite different from those found in forests today. Their similarity to the trees of northern Pakistan at the time is a reminder of the close and very different neighbours Madagascar had then. Fossils from other regions of the world suggest that Madagascar's forests were home to ancient amphibians and reptiles, although no trace of them has yet been found.

This faintly glimpsed early world largely disappeared in a global wave of extinctions around 252 million years ago. It was the third of five waves that swept Earth long before our own appearance on the scene. The first, about 439 million years ago, happened soon after plants began colonising the land and well before land animals evolved. Global cooling seems to have been the cause, producing a shock great enough to wipe out many marine creatures. A second, similar episode a hundred million years or so later likely caused another major die-off.

The third was a truly cataclysmic upheaval. The likely cause this time was a rock from outer space that hit Earth and set off paroxysms of volcanic activity. Fire, acid rain, plunging oxygen levels, and wide fluctuations in atmospheric carbon dioxide abruptly eliminated many forms of life on land and in the sea. Fungi briefly flooded the world, nourished by a litter of dead and decaying matter. Scientific articles about these events read like the screenplay for an environmental horror film. Preserved pollen and spores discovered in Madagascar provide clues to what happened there, and hint at a whole new world of vegetation. The remains contain no traces of the ginkgo- and pine-like trees from earlier times. Ferns and club mosses had replaced them. Living club mosses are small, but some species grew to be huge trees in the past, and a vision of strange club moss forests covering Madagascar's landscapes floats into sight.

Amphibians, reptiles and fish dominated Earth in this long-ago epoch, variously scampering, plodding, hopping, slithering or swimming about their business, and the Morondava Basin in western Madagascar offered near-perfect conditions to become a repository of the fossil remains of species that survived the mayhem of the third mass extinction. Great rivers carried bones and carcasses downstream to lowland basins, with plenty of sand and mud to bury them while they turned into fossils, and there was little volcanic or mountain-building activity in the region to obliterate them during the millions of years that followed.

Repositories are one thing, finding them and figuring out what you have found a second. We have gritty teams of palaeontologists to thank for this, who search until they 'hit gold' and then pursue the arduous task of identifying fragments of fossilised bone in order to unlock Madagascar's ancient past. John Flynn and Lovasoa Ranivoharimanana led the team whose discoveries I describe first. The fieldwork took place in the early 1990s, and I met John many years later. He was Dean of the Gilder School at the American Museum of Natural History by then, and I had difficulty picturing the distinguished figure in a smart suit with a spade in his hand – until he began talking about his experiences in the field! Sifting through deposits in the Morondava Basin over 200 million years old, the team discovered animals of kinds unlike any seen anywhere in the world today. Fossil gold . . .

The first finds were fragments of the jaw of a rhynchosaur. These goat-sized, parrot-beaked relatives of dinosaurs are quite common in rocks of about the same age in other parts of the world. Hot on the heels of the rhynchosaur discovery came the 6-inch-long skull of a traversodontid, neither reptile nor mammal, though on the line leading to mammals. At least four species of these odd creatures have been identified, ranging in size from goats to alligators. Rhynchosaurs and traversodontids lived on the ground, walking on all fours. Teeth are good indicators of diet, and the preserved teeth of these animals suggest they ate plants, perhaps grazing and browsing together in herds like – and totally unlike – wildebeest and zebra on the grasslands of East Africa today.

The rhynchosaurs and traversodontids had company. Alongside them roamed archosauromorphs, whose skulls and teeth indicate that they too chomped on plant foods. Once thought to be true dinosaurs, it is now clear that they were at the root of a broad group including dinosaurs and several other reptile forms. Their skulls and teeth indicate that they too had

a plant diet. But not everyone in the community was a vegetarian. Small creatures called chiniquodontids had sharp, pointy
teeth signalling that they were little carnivores. Their other claim
to fame is that they may be more closely related to living
mammals than any other creature living at that time, traversodontids included. Spotting the defining characteristics of a
mammal is pretty straightforward among living species. Mammals
are warm blooded, and have hairy or furry bodies. Females
produce milk to suckle their young, and give birth to live young
(except for five species that lay eggs, most famously the platypus).
Fossils preserve none of these features, of course. Identifying
mammals from teeth, skulls or skeletons alone is difficult, and
ancient species with a mosaic of features also pose a dilemma.
Animals 'on the line leading to mammals' or 'closely related to
living mammals' are good reminders of this: assigning the
natural world to clearly labelled pigeonholes creates boundaries
that often do not actually exist in nature.

*John Flynn (third from right) and Lovasoa Ranivoharimanana
(right) at work with the team (AMNH/J. Flynn)*

The strange, distant community of the Morondava Basin recently expanded with the discovery of a fifth kind of animal – in the museum drawers where all the fossils and rocks collected in the 1990s were safely stored for further study. The team gave it the scientific name *Kongonaphon kely*. *Kongona* is a word for bug in the Malagasy language, and *kely* means small or tiny. Throw in a bit of Greek, and the formal name is translatable into common English as 'tiny bug slayer'. And what was this bug slayer? A dinosaur, just 10 centimetres tall. Accustomed as we are to the ferocious dinosaurs of *Jurassic Park*, there is something particularly endearing about this little creature. Vicious it was not, at least not in human terms. Pits in its conically-shaped teeth suggest that it ate insects – hence the name.

What happened to all these species? Perhaps some succumbed in the wake of a convulsion of volcanic activity accompanied by global warming that triggered a fourth wave of extinctions around 200 million years ago. We can only speculate, because no geological deposits containing a record of what was happening in Madagascar during this period have been discovered. As a general matter, this is not surprising. The fate of most deposits is to be buried by further deposits, putting them beyond the reach of eager palaeontological eyes millions of years later. Even if exposed, moreover, deposits are often empty of fossils because conditions for the preservation of animal bones were poor.

Moving northward to the Mahajanga Basin near the modern city of Mahajanga, we come to the next finds by Flynn and his team. These discoveries were younger, only about 167 million years old, and they included dinosaur teeth, fish scales and unidentifiable bone fragments. Mixed in with the jumble of remains, though, the team spotted a partial jawbone the size of a rice grain, with three teeth in place. Without a doubt, the jaw belonged to a tiny, primitive mammal. The original owner, probably no bigger than a modern shrew or mouse lemur, was firmly classified as a mammal and given the scientific name *Ambondro mahabo*.

Plants lost out in the palaeontological sweepstakes of preservation and discovery at these sites in the west and northwest. In the absence of plant remains, the vegetation amid which animals lived remains a matter of conjecture. A mix of club mosses, ferns and other kinds of ground cover likely predominated in the humid conditions that prevailed, and the presence of animals that walked on all fours and ate plants certainly suggests that there was plenty of vegetation within reach on the ground.

The fossils from the northwest offer our last glimpse of Madagascar when it was still locked inside Gondwana, and their discovery has also changed ideas about the course of evolution globally. Take *Ambondro*, for example. It was interesting to find even a small fragment of a mammal in Madagascar so early, demonstrating that mammals lived there alongside dinosaurs for a hundred million years or more. But interest does not end there. Diminutive, rare, and lacking glamour, mammals had gone undetected at such an early time anywhere in Gondwana up to this discovery. *Ambondro* pushes back the earliest firm 'geological sighting' by millions of years, and shows for the first time that primitive mammals were present in Gondwana as well as northern lands.

*

Spinning the clock's hands forward brings us to further discoveries in the Mahajanga Basin, and more fossil gold! These were found in deposits roughly 72–66 million years old, laid down as the age of dinosaurs was nearing an end. By now, Madagascar was a low, flat island, blocked from tropical currents by India and still south of its present location, well out of the path of tropical cyclones that hit the island today.

David Krause, another gritty palaeontologist, was in the lead this time. He and his team pieced together clues to the area's climate and vegetation from ancient soils and a spectacular array of animal fossils, because there were no plant fossils to be found. It was a balmy landscape overall, but huge accumulations of fossilised bones inject a dark note into the picture. A 'mass killer' was lurking somewhere. That killer was likely drought. Carefully excavated soils contained well-preserved traces of deep roots, common today in plants that reach far into the ground for moisture and nutrients in order to cope with dry conditions. Although a green and pleasant land of verdant forest and wetlands much of the time, the northwest must have been a tough place to survive in some years.

The reptiles, amphibians and fish occupying this landscape were very different from those that lived in the region a hundred million years earlier. They were not one bit like the modern wildlife of Madagascar either. Among other things, several were fearsome. Consider the theropod (literally, 'beast-footed') dinosaurs *Masiakasaurus knopfleri* and *Majungasaurus crenatissimus*. Both had small forelimbs and big strong hindlimbs. Both were predators, one small and agile, the other big and ponderous. Less than 2 metres long, *Masiakasaurus* likely made up for its small size by the speed at which it moved and chased down prey. The genus was well named: *masiaka* in Malagasy roughly translates as 'fierce'.

Majungasaurus, on the other hand, could use a more appropriate name than that of the modern city nearby. Six or seven metres long – a lot bigger than *Masiakasuarus* – it may have weighed more than 1,000 kg. *Majungasaurus* was the top predator in the region, with a big appetite and undiscriminating tastes. Many fossil bones mixed in with the dinosaur remains bore marks that could be traced to the teeth of *Majungasaurus*. The bones came from a wide range of species, and from individuals that were young, old, large-bodied or small. In contrast to other meat-eating dinosaurs, which typically targeted particular types of prey, *Majungasaurus* was clearly not a picky feeder. It was probably a scavenger at least some of the time, feeding on dead or dying animals – perhaps during a drought. And some *Majungasaurus* bones displayed tooth marks of . . . *Majungasaurus*. These dinosaurs had such broad tastes they even ate each other – the only theropods with demonstrated cannibal tendencies.

But not everyone was fierce. The real giants of the Mahajanga Basin, two species of sauropods, were dinosaurs of the benign variety. Members of the larger species were perhaps 10 metres from nose to tail, weighing as much or more than a modern elephant but built very differently, with a long neck and tiny head balanced by a long tail. They plodded around on all fours, lived in social groups, and browsed on plants.

In the sky above the earthbound critters flew birds, and perhaps one species of flying dinosaur. Birds, ancient and modern, evolved from dinosaurs, and the question of when a dinosaur is actually a bird, and *vice versa*, remains much debated. At least six species were represented among the abundant bird, or near-bird, remains, the largest the size of red-tailed hawks, the smallest no bigger than tree sparrows. But none are ancestral to modern birds. Remains of definite ancestors of modern birds show up in similarly aged deposits in other parts of the world, but what we see in the Mahajanga Basin are early experiments in birdiness.

Snakes equalled these experimental birds in diversity, with six species including the largest snake anywhere in the world at the time, *Madtsoia madagascariensis*. Almost 8 metres long and probably heavy-bodied, it likely sat and waited to strangle unsuspecting prey that happened by. Crocodilians were diverse too, with at least seven species, and they include my favourite reptile of the period. Known from a well-preserved fossil skull, the scientific name of this crocodile is *Simosuchus clarki*. *Simosuchus* had nostrils and eye sockets on the side. This suggests it was not well adapted for floating at the surface of the water as modern crocodiles do, with nostrils and eyes on top so they can see and breathe while they float. It departed from general notions of crocodile-hood in other ways too. Modern crocodiles have cone-shaped teeth in a long snout, with specialised jaw joints and a flattened skull – all geared to delivering an exceedingly powerful bite. *Simosuchus* was a friendly critter by comparison, with a short, blunt snout and teeth reminiscent of plant-eating lizards like some dinosaurs and, today, green iguanas. Vegetarian crocodiles? Why not . . .

Frogs, lizards and turtles round out the amphibian and reptile fauna. The frogs deserve a special word. Only one species has been identified and described so far, with the splendid scientific name of *Beelzebufo ampinga*. The name by which its discoverers evidently referred to it is even better – the Frog from Hell. This was no garden-variety amphibian. The Frog from Hell was huge, 42 cm long with a skull 20 cm wide – picture a 5 kg house cat – and would dwarf the largest frogs alive anywhere in the world today.

Beelzebufo left no living descendants in Madagascar. Its closest living relative is the horned frog in South America. These carnivorous frogs with sharp teeth and a powerful bite have been dubbed 'hopping heads', and the bite force of *Beelzebufo*, with even bigger and more powerful jaws, is estimated to have been greater than that of a crocodile. Adults would have been terrifyingly efficient predators on small vertebrates and, who knows, hatchling dinosaurs. Vegetarian crocodiles *and* giant, dinosaur-munching frogs? Why not . . .

Finding fossils of any kind involves great expertise, effort, luck – and persistence. Finding evidence of early mammals is especially hard, because they were small and probably rare too, making the search akin to looking for tiny needles in a huge haystack. In over 20 years spent looking for direct descendants of *Ambondro* or its mammalian relatives, Krause and his team's only reward was a handful of tooth fragments, distinctive enough to know they came from mammals but too small and incomplete to identify the species. Compare that to over 15,000 specimens of reptiles and amphibians.

Better luck came with the discovery of a single, intact cheek tooth that could be identified as that of a marsupial mammal. For the first time, this put marsupial mammals on the map in Madagascar 68 million years ago and, as with *Ambondro*, the discovery told a bigger story too. Marsupials probably originated in the northern hemisphere and made their way south to Gondwana. The Mahajanga tooth suggests they became widespread there, making their way to Madagascar from South America via Antarctica. But this global success story, if such it is, did not last. Today, marsupials are found only in Australia, South America and parts of North America.

Still, a single tooth can only tell you so much and a whole skull much, much more. A skull discovered in the Mahajanga Basin in 2010 makes this clear. Given the name *Vintana sertichi*, at the time it was one of only three complete mammal skulls from all of Gondwana during this period and was precious indeed. It exhibits a hodgepodge of primitive mammalian features, non-mammalian features, and features all its own. There is nothing like *Vintana* in Madagascar or anywhere else in the world today, and it is not a candidate ancestor of any of the living mammals. Like almost all of the animals in this chapter, it belonged to a different, earlier world.

What does this skull tell us about the life of its long-ago owner? Judging from the skull's size, *Vintana* weighed almost 10 kg, about as big as a small beagle. The eye sockets were large and, from the shape of the braincase imprinted on the inside of the skull, so were the olfactory lobes of the brain. The cheek teeth were big and, in this specimen, heavily worn; the incisor teeth were long, curved and probably continuously growing like those of a rodent. Taken together, these and other clues point to a nimble creature with a diet of tough or abrasive foods – like seeds, hard-shelled fruit and roots. The bony structure of its ear hints at a capacity to hear sounds at exceptionally high frequencies, and perhaps *Vintana* snagged insects on the wing as well.

And then there's *Adalatherium hui*. It is tempting to explain the meaning of the scientific names given to many of Madagascar's animals and plants because they are often little histories unto themselves, embedding the names of researchers, and reactions and emotions at the time of discovery. I try to resist the temptation, but fail (again) in this instance. *Hui* pays homage to a great palaeontologist who died too young. *Adala* means 'crazy' in Malagasy – an adjective attached to me from time to time by laughing friends – and *therium* is Greek for 'beast'.

Crazy beast. It certainly says something about this creature, known from a complete skeleton, not just a skull like *Vintana*. Study of the skeleton revealed a young individual, about a third the size of *Vintana*. Weighing around 3 kg, it was still very large for a mammal at the time – most were mouse-sized. *Adalatherium* walked on all fours on the ground. So far, this all sounds reasonably normal. Details of its anatomy are 'crazy', however – unlike any other mammal alive then – and likely due to evolving in island isolation. What did *Adalatherium* do for a living? Krause and his colleagues are on the case, but as yet there is no answer.

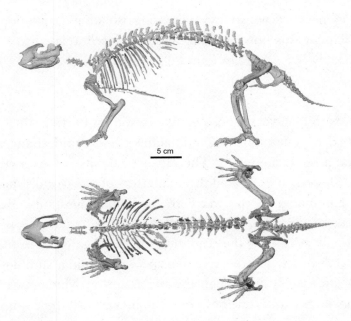

The exceptionally intact skeleton of Adalatherium hui
(Hoffmann et al. 2020)

The closest relatives of the Mahajanga Basin mammals, reptiles and frogs lived in India and South America, and all these species, except for one dinosaur, were earthbound creatures. These geographical connections suggest that their ancestors were 'on board' millions of years, perhaps tens of millions of years, before Madagascar was surrounded by water. Almost certainly, they originated in Madagascar when it was part of Gondwana, or in another region of Gondwana and made their way overland to the area that would become an island.

The Gondwana origins of Madagascar's pre-asteroid fauna stretch deep into the past. By 70 million years ago, however, wildlife in the Mahajanga Basin had become a Gondwana fauna-with-a-difference. While still retaining the distinctive stamp of their ancestry, species were evolving and developing their own character in the increasing isolation of island life. Madagascar was now home to animals closely related to – but not exactly like – species found elsewhere.

*

Sixty-six million years ago, a giant asteroid hit Earth in the Gulf of Mexico, unleashing volcanic activity and rapid changes in climate around the world. The fifth global wave of extinctions followed, and an estimated three-quarters of the world's species were eliminated in the mayhem, including dinosaurs. But a number of mammal species – all rat-sized or smaller – made it through, heralding the beginning of the 'Age of Mammals'. And what of the elusive mammal species scuttling, climbing or threading their way through the vegetation of Madagascar 66 million years ago? None have descendants among the mammals living there today. None. They too were swept away.

The animals that starred in this chapter have a bittersweet quality. They live on in the form of fossils and tell of long-ago exotic faunas, reminding us that life in the past was profoundly and intriguingly different from the present. At the same time, they stand in poignant contrast to the *lack* of fossils to illuminate the next phase in Madagascar's history. Like the period around 200 million years ago, there are no exposed, fossil-bearing deposits. This is particularly frustrating because it spans the time during which the island's modern fauna evolved. Ending this chapter 66 million years ago in our imagination, the island is all but empty of animals. What happens? How is Madagascar to acquire a new array of wildlife? In the absence of fossils, researchers have found other ways to push the window partway open on to this period in the island's history.

The western tree boa, Sanzinia volontany
(Robert Harding/Alamy Stock Photo)

CHAPTER 4

Starting Over

Adrama in the forest at Ampijoroa way back in 1970 first made me wonder how and when ancestors of the animals living in Madagascar today arrived. Time was going by slowly that day, and I was ready for a lunch break of cold rice and sardines in the shade of a tree. But one *sifaka* was still munching away on a fistful of leaves when the rest of her group moved out of sight, and I wanted to keep an eye on her until she rejoined the others and settled down for a midday siesta. She ate, and I sat hungrily, until a sudden explosion of alarm calls banished all thoughts of food from both our minds. She leapt off through the trees, and I hurried in her wake.

The scene we came upon was disconcerting. The other members of her group were arrayed in a circle, staring silently. The object of their fascination was a boa snake in a tree with a bulging-eyed mouse lemur slowly disappearing feet-first down its throat. My first thought was 'why don't they *do* something?'. 'Should *I* do something?' followed quickly. It is not easy to witness nature red in tooth and claw without intervening. I doubt the *sifaka* shared my existential dilemma, but the event was clearly a matter of interest to them and we all kept watching until the mouse lemur vanished from sight and reappeared as a lump in the boa's sleek body.

The episode made the day exciting in a gruesome kind of way. My research usually involved watching lemurs do nothing at great length, and idle thoughts would often float through my mind at those times. But that day set me wondering if there were lemurs in Madagascar before it became an island, and about the arrival time of animals that consider lemurs a potential lunch. Did that boa's ancestors live in Madagascar before lemurs were around, or were the lemurs horribly shocked when boas appeared on the scene? I had no answer to this question then and, in fact, little evidence bearing on them existed at the time.

Recent feats of molecular wizardry have established that the ancestors of almost all animals alive today arrived after Madagascar became an island, and after the asteroid hit Earth 66 million years ago. In a quirk of geological history, deposits laid down during this period of arrivals are rarely exposed to human eyes and the few discovered so far are heavily weathered and virtually empty of fossils. As a result, fossil clues to the course of events when the island's living plants and animals evolved are rare indeed. The record resumes only about 150,000 years ago, and most remains are less than 10,000 years old. These remains, discovered in recent deposits, are sub-fossils, in which the process whereby minerals replace organic matter in bone is not yet complete and genetic material is sometimes still present.

Researchers keep looking for fossil deposits, but for now the big gap in the record makes the task of evolutionary reconstruction particularly difficult and uncertain. In the last few decades, however, molecular biology has opened up a new avenue to the past by reverse-engineering from the present. Comparing the molecular profiles of living species and sub-fossil remains makes it possible to establish the closeness of their evolutionary relationships, trace their lineages back into the past, and estimate when a lineage's founding ancestor reached Madagascar. A lineage is composed of animals or plants descended from a common ancestor. The term is used with varying levels of inclusiveness, unlike categories in the formal scientific naming system. This system classifies different forms of life in a clear hierarchy: a *genus* groups together closely-related species, a *family* or *sub-family* groups together closely-related genera, and an *order* groups together families. It is a good and useful way to organise the natural world, even though it can be difficult to figure out where species fit or even if they fit into the system at all.

Scientists call a lineage for which there is no fossil record a ghost lineage. Reconstructing and dating ghosts using molecular techniques is a complicated business. The first step is to estimate the genetic distances between pairs of related species alive today; the second is to make a judgment about the rate at which their genetic material accumulated random molecular changes, or mutations; the third is to infer when they last shared a common ancestor, using the rate of random mutations as a molecular clock. Molecular clocks are calibrated against fossils of known dates whenever possible, but the ghost lineages of Madagascar are necessarily products of molecular estimate and inference alone – and particularly ghostly as a result.

The time at which a ghost lineage split off from its nearest relatives in another land is assumed to be roughly the time when the lineage's founders arrived in Madagascar. Estimates usually span several million years, even with the most sophisticated techniques, and it is hard to keep a single year, let alone a single day in mind at the same time as an estimate spanning millions of years. And that makes it easy to forget we are considering the history of individual animals arriving on a particular day or night, scrambling ashore hungry, thirsty and wet, or plopping down wearily on a tree limb after a long flight. The immediacy of life is in constant need of rescue as we plunge into the past where time can only be measured in millions of years.

Because almost everything we know about these events is by reverse-engineering from the present, this chapter is also an introduction to the island's land-living amphibians, reptiles, mammals and birds. The table on the next page lists the families in these groups along with the number of firmly established and named species in each, and indicates which are endemic. It includes only species found in protected areas, but likely represents all families. Excellent, comprehensive field guides to much of Madagascar's wildlife exist and I do not try to replicate them here. My focus is on the broad sweep of history and the evolutionary patterns that can be discerned. Before digging into these patterns, however, let us consider how any land animals at all managed to cross a stretch of open sea 400 kilometres wide or more, and the challenges they faced along the way.

*

Madagascar was clearly in a fix after the asteroid crisis. Surrounded by deep water now, how could the island replenish its wildlife? The question stumped early natural historians puzzling over similarities between the island's plants and animals and those in Africa, India, and even South America and Australia. Lemuria, the sunken continent linking Madagascar to these places according to some nineteenth-century natural historians, provided one possible answer but the geological evidence was sparse and equivocal. Researchers long viewed

Amphibians (341 spp.)	Reptiles (409 spp.)	Mammals (226 spp.)	Birds (253 spp.)
Frogs	**Lizards**	**Primates**	Accipitridae (15 spp.)
Bufonidae (1 spp.)	Chamaeleonidae (**85 spp.)	*Cheirogaleidae* (**42 spp.)	Acrocephalidae (3 spp.)
Dicroglossidae (1 spp.)	Gekkonidae (*111 spp.)	*Daubentoniidae* (**1 spp.)	Alaudidae (**1 spp.)
Hyperoliidae (**11 spp.)	Gerrhosauridae (*19 spp.)	*Indriidae* (**19 spp.)	Alcedinidae (2 spp.)
Mantellidae (**215 spp.)	Opluridae (8 spp.)	*Lemuridae* (**21 spp.)	Anatidae (10 spp.)
Microhylidae (**112 spp.)	Scincidae (**77 spp.)	*Lepilemuridae* (**26 spp.)	Anhingidae (1 spp.)
Ptychadenidae (**1 spp.)	**Turtles and Tortoises**	**Tenrecs**	Apodidae (5 spp.)
	Pelomedusidae (3 spp.)	*Tenrecidae* (**32 spp.)	Ardeidae (14 spp.)
	Podocnemididae (**1 spp.)	**Carnivores**	*Bernieridae* (**11 spp.)
	Testudinidae (5 spp.)	*Eupleridae* (**9 spp.)	*Brachypteraciidae* (**5 spp.)
	Crocodiles	**Rodents**	Campephagidae (**1 spp.)
	Crocodylidae (1 spp.)	Muridae (2 spp.)	Caprimulgidae (2 spp.)
		Nesomyidae (**28 spp.)	Charadriidae (9 spp.)
	Snakes		Ciconiidae (2 spp.)
	Boidae (**4 spp.)	**Bats**	Cisticolidae (4 spp.)
	Lamprohiidae (**82 spp.)	Emballonuridae (4 spp.)	Columbidae (4 spp.)
	Typhlopidae (12 spp.)	Hipposideridae (**2 spp.)	Coraciidae (1 spp.)
	Xenotyphlopidae (**1 spp.)	Miniopteridae (12 spp.)	Corvidae (2 spp.)
		Molossidae (8 spp.)	Cuculidae (12 spp.)
		Myzopodidae (**2 spp.)	Dicruridae (1 spp.)
		Nycteridae (**1 spp.)	Dromadidae (1 spp.)
		Pteropodidae (**3 spp.)	Estrildidae (2 spp.)
		Rhinonycteridae (**3 spp.)	Eurylaimidae (**4 spp.)
		Vespertilionidae (11 spp.)	Falconidae (5 spp.)
			Fregatidae (1 spp.)
			Glareolidae (**1 spp.)
			Hirundinidae (5 spp.)
			Jacanidae (**1 spp.)
			Laridae (2 spp.)
			Leptosomidae (1 spp.)
			Locustellidae (**2 spp.)
			Meropidae (2 spp.)
			Mesitornithidae (**3 spp.)
			Monarchidae (1 spp.)
			Motacillidae (**1 spp.)
			Nectariniidae (2 spp.)
			Numididae (1 spp.)
			Oriolidae (1. spp.)
			Pelecanidae (1 spp.)
			Continued overleaf

Living families of land vertebrates, with species numbers in parentheses;
double asterisk signifies all species present in Madagascar are endemic, single
asterisk at least 95 per cent; endemic families are shown in italics (data from
Goodman et al. eds., Vol. 1, 2018)

Amphibians (341 spp.)	Reptiles (409 spp.)	Mammals (226 spp.)	Birds (253 spp.)
			Phaethontidae (1 spp.)
			Phalacrocoracidae (1 spp.)
			Phasianidae (3 spp.)
			Phoenicopteridae (2 spp.)
			Ploceidae (5 spp.)
			Podicipedidae (2 spp.)
			Psittacidae (3 spp.)
			Pteroclidae (**1 spp.)
			Pycnonotidae (1 spp.)
			Rallidae (12 spp.)
			Recurvirostridae (2 spp.)
			Rostratulidae (1 spp.)
			Scolopacidae (18 spp.)
			Scopidae (1 spp.)
			Sternidae (16 spp.)
			Strigidae (4 spp.)
			Sturnidae (3 spp.)
			Threskiornithidae (4 spp.)
			Turdidae (5 spp.)
			Turnicidae (**1 spp.)
			Tytonidae (2 spp.)
			Upupidae (**1 spp.)
			Vangidae (*21 spp.)
			Zosteropidae (1 spp.)

with scepticism the idea that land animals could reach distant lands by sea. But not all did: 'Any event that is not absolutely impossible . . . becomes probable if enough time elapses' commented the late, great palaeontologist George Gaylord Simpson over 60 years ago, when he proposed a 'sweepstakes route' for mammals across the Mozambique Channel from Africa. Decades were to pass before his idea became widely accepted.

Today, most researchers agree that except for a few survivors of the asteroid's impact, and birds and bats that could travel by air, everyone reached Madagascar by sea from Africa across the Mozambique Channel. (If new hints of the intermittent presence of a land bridge are confirmed, however, this chapter will need rewriting!) A few probably swam or floated across, but most came by rafting aboard mats of vegetation. Many kinds of animals never made the crossing successfully or, if they did, died out without leaving living descendants. In the absence of a fossil record, of these there is no trace. Those that survived to the present are overwhelmingly endemic, a term applied to species found exclusively in one place. In practice, some species spread from Madagascar to nearby islands and are not endemic to Madagascar in the strict sense, but I refer to them that way to keep matters as simple as possible (which, of course, they are not).

Direct observations of sea journeys by animals that are not strong swimmers are rare, but there is good evidence from around the world that they happen, and also how they happen. A mat of tree trunks, vines and earth torn from a riverbank carries animals out to sea, and ocean currents then propel the mat-turned-raft to another shore. The Zambezi is the largest river along the African coast flowing into the Mozambique Channel; it lies due west of Madagascar, and mats of vegetation drifting from its mouth are a common sight. Those are the positives. But the river flows into one of the widest stretches of the Channel, and floating all the way to Madagascar would be a longshot at the best of times.

The 'best of times' were the tens of millions of years when sea currents flowed west across the Indian Ocean, hit the coast of Africa, and bounced back eastward toward Madagascar. With estimated speeds of more than 20 centimetres per second during the cyclone season, it could have taken less than four weeks for a mat to drift across the Channel. This watery umbilical cord was cut when the predominant direction of currents changed about 23 million years ago, and the chances of making the crossing eastward became slimmer still.

Even with favourable currents, the voyage was a long haul for animals with little or no food and no fresh water except occasional rains. How did they do it? The magnitude of the challenge differed for reptiles, amphibians and mammals. Reptiles have a relatively low energy turnover, their body temperature tracks external temperatures, and some can go for months without food or water. Some, notably chameleons, lay eggs that can remain in a state of suspended development for many months at a wide range of temperatures. Conceivably, reptilian eggs rather than animals made the voyage, bobbing over the waves. Either way, there are many instances of lizards and snakes crossing wide oceans. For amphibians it is much more difficult, because a sea journey is a particularly dangerous proposition: their skin is permeable to water and they dry out quickly, and salt water dehydrates them even faster.

Mammals face another kind of difficulty, with their high energy turnover, high and stable body temperature and constant need for food to stay alive, but they too survive journeys that defy the imagination. Voyages across the Atlantic Ocean by the ancestors of New World rodents and monkeys are the most astonishing of all, perhaps accomplished by using islands that existed at the time as stepping-stones.

The ancestors of Madagascar's living mammals may have had their own particular way of coping. Each of the endemic groups on the island today includes species or close relatives in Africa capable of lowering their energy requirements and body temperature quite drastically when food is scarce. Some use this ability to hibernate or aestivate (like hibernating except it happens in hot, dry conditions rather than wintertime). These species store fat reserves to tide them over for weeks or months without food. Others go into torpor, and just get sleepy when need arises. Torpor is shorter (hours or days) and more opportunistic, used in emergencies for 'snoozing through disaster'.

Cheirogaleus crossleyi, *descendant of animals that likely snoozed across the Mozambique Channel (photograph by Ken Behrens)*

The idea that hibernation or torpor could have helped the island's mammalian ancestors survive the crossing was seeded 30 years ago and has taken hold since then. But getting to Madagascar was only a first step. The future depended on producing offspring. The ancestor of a lineage could possibly have been a single pregnant female. More probable, 'the ancestor' was a handful of animals – a female with young, or males and females that made landfall together. Whatever the number of individuals involved, success was tantamount to winning the jackpot, and a sad vision comes to mind of countless drowned bodies slowly sinking in the sea, coupled with the rarest of events – survival.

*

Frogs are the only amphibians of the three orders in the world that are known to have reached Madagascar, reflecting the deadly perils of the voyage for creatures of this kind. They were among the earliest vertebrates to arrive by sea, around 60–58 million years ago. All but two species of living frogs are endemic, with a further 200 or so 'candidate' species that are not yet firmly established or formally named. Only Brazil, Colombia and Mexico have more. Almost all Malagasy frogs belong to just two families.

A widespread and common frog today, *Ptychadena mascariensis*, stands out from this pattern. Its ancestor arrived in the not-too-distant past, even though currents had long flowed predominantly away from Madagascar toward Africa by then. In fact, Madagascar's *P. mascariensis* looks so much like its African relative that it was long thought to be the same species, brought over by people in the last few thousand years. This turns out not to be so. Separation of the Malagasy frog from its African relatives happened thousands of years before people arrived. The rarest of events provides a likely explanation. Very occasionally, cyclones barrelling at the African coastline from the Indian Ocean ricochet back eastward across the Mozambique Channel and temporarily shift currents in that direction too. These fast-flowing currents sweep mats of vegetation along with them, and apparently one carried a handful of small, puzzled frogs to their new home. *P. mascariensis* is proof that every rule has exceptions.

Madagascar's living reptiles include a different kind of exception to the rule – iguanas and blind snakes, the only land species with ancestors that almost certainly date from pre-asteroid times. One wonders why their lineages persisted. Perhaps it was evolutionary luck, or something yet to be determined about their biology and behaviour that helped them survive the crisis and subsequent vicissitudes of life in Madagascar. Ancestors of all other living reptiles arrived by sea. With 13 families today, just one of them endemic, reptiles evidently fared a lot better than amphibians in the Channel sweepstakes.

The seafarer's prize surely goes to lizards – Madagascar's chameleons, geckos, skinks and plated lizards. Crossing the Channel was a challenge even for accomplished sailors and there are notable absences, but lizards certainly got around. Arriving from Africa on multiple voyages, some then rafted off to nearby islands while others – according to the genetic evidence – seem to have gone back and forth between Madagascar and Africa. The hubbub of comings and goings makes it even more difficult than usual to figure out what happened and when, but diversity and high levels of endemism are both clearly on display. Altogether, Madagascar has around 409 living reptile species, with many additional candidates – and most of them are lizards.

Take chameleons. Almost half the world's chameleons are unique to Madagascar, with around 85 endemic species. They probably came from Africa, with one arrival around 65 million years ago and another about 47 million years ago. Today, Madagascar's chameleons range from monsters over half a metre long, the biggest in the world, to the tiny creatures that fit on your thumb print – not just the smallest chameleons in the world but one of the smallest reptiles of any kind anywhere. I share the fascination of the Malagasy with these creatures. They melt out of sight in the cool dry season at Bezà Mahafaly, but in the warmth of summer whole trees fill with them stalking slowly about their mysterious business. A foot poised in mid-air, one eye swivelled back, the other forward, long tongue shooting out to lick an insect off a branch . . .

Land tortoises, possibly the world's most ancient group of living reptiles, spread widely across many continents between 55 and 35 million years ago. They were latecomers to Madagascar though, arriving from Africa sometime between 22 and 11 million years ago. Maybe they rafted across on a cyclone-driven current, but they are also the best and perhaps only candidates to have floated over.

Tortoises do float. I learned this firsthand, or almost firsthand. In late January 2005, two cyclones hit southwest Madagascar in the space of four days. Seventy-eight people died, and more than 32,000 were left homeless. No one died at Bezà Mahafaly but there was the biggest flood in living memory, and a metre of water engulfed our research camp. A heavy, undulating layer of silt and sand covered much of the forest floor between camp and the river a half-mile away, and the churning waters washed away big chunks of farmland on the far bank. Visiting soon afterwards, we surveyed the devastation with horror. Farmers' crops had been swept away or ruined, and we were certain that the forest's many tortoises (*Astrochelys radiata*) had drowned in the rising water. We were wrong. Individuals with carapaces we had marked before the flood, in order to monitor them over the long-term, resumed their plodding along the forest floor after the water receded. Buoyancy saves lives!

Starting Over

Mammals

family
(Primates)

Archaeolemuridae   Archaeolemur edwardsi
                  Archaeolemur majori
                  Hadropithecus stenognathus

Megaladapidae   Megaladapis edwardsi
                  Megaladapis grandidieri
                  Megaladapis madagascariensis

Palaeopropithecidae   Archaeoindris fontoynontii
                  Babakotia radofilai
                  Mesopropithecus dolichobrachion
                  Mesopropithecus globiceps
                  Mesopropithecus pithecoides
                  Palaeopropithecus ingens
                  Palaeopropithecus kelyus
                  Palaeopropithecus maximus

Daubentoniidae   Daubentonia robusta

Lemuridae   Pachylemur insignis
                  Pachylemur jullyi

(Tenrecs)
Tenrecidae   Microgale macpheei
(Carnivores)
Eupleridae   Cryptoprocta spelea
(Rodents)
Nesomyidae   Brachytarsomys mahajambaensis
                  Hypogeomys australis
                  Nesomys narindaensis

(Bats)
Hipposideridae   Hipposideros besaoka
                  Triaenops goodmani

(Hippopotamus)
Hippopotamidae   Hippopotamus laloumena
                  Hippopotamus lemurlei
                  Hippopotamus madagascariensis

Order Bibymalagasia   Plesiorycteropus germainepetterae
                  Plesiorycteropus madagascariensis

Reptiles

family
(Crocodiles)
Crocodylidae   Voay robustus
(Tortoises)
Testudinidae   Aldabrachelys abrupta
                  Aldabrachelys grandidieri

Birds

family
Accipitridae   Stephanoaetus mahery

Aepyornithidae   Aepyornis hildebrandti
                  Aepyornis maximus
                  Mullerornis modestus
                  Vorombe titan

Anatidae   Alopochen sirabensis
                  Centrornis majori

Brachypteraciidae   Brachypteracias langrandi

Charadriidae   Vanellus madagascariensis

Cuculidae   Coua berthae
                  Coua primavea

Rallidae   Hovacrex roberti

Recently extinct species of land vertebrates; extinct families are in italics (from Goodman & Jungers 2014, and Hansford & Turvey 2018 for Aepyornithicidae)

77

Six endemic tortoise species evolved from their floating or rafting ancestor, including one of the world's smallest – and two of the largest, both now extinct. Giant tortoises evolved independently on most continents, perhaps as a way of escaping predators. Madagascar's giant tortoises (*Aldabrachelys spp.*) went on to colonise other islands in the western Indian Ocean, although the only remaining wild population is on Aldabra. Human hands were at work in some cases, perhaps early sailors establishing provisioning stations. But floating or rafting scenarios may apply to others. The sheer size of giant tortoises – topping the scales at over 200 kg – does not bring buoyancy immediately to mind, but in fact, like smaller tortoises, they float well. One washed ashore on the coast of Tanzania in 2004. Alive if a bit the worse for wear, it had evidently drifted for several weeks and probably came from Aldabra, some 740 km away. If so, giant tortoises surely qualify as the champions of floating.

Madagascar's crocodiles (*Crocodylus niloticus*) are the island's largest living reptiles and most fearsome of any animal there today. They are genetically indistinguishable from their African relatives the Nile crocodiles, and probably swam to Madagascar in the last few thousand years. *Voay robustus*, even bigger (though Madagascar's modern crocodile is not small), diverged from Nile crocodiles far in the past and likely reached Madagascar many million years ago. Toothmarks on bones show that these crocs hunted the largest-bodied lemurs, including adults. Today, sub-fossils are the only record of their existence. Perhaps they died out when their main prey went extinct or perhaps newly arrived Nile crocs outcompeted them. Perhaps people killed these dangerous animals when they could.

The island's snakes are a friendlier bunch. Venomous adders, cobras, mambas and vipers are absent – another good reason to work in Madagascar. Most Malagasy snakes belong to the world's largest snake family, the Lamprophiidae, and all 75 or so species appear to trace back to an ancestor that arrived from Africa less than 35 million years ago. This brings me back to the ancestor of that boa (*Sanzinia volontany*) swallowing a mouse lemur in the forest at Ampijoroa. When and how did it arrive? The evidence is mixed. Snakes that strangle their prey occur in many parts of the world, and molecular and morphological similarities among them suggest that some, including Madagascar's boas, share a common ancestor dating back to Gondwana times. The boa-like snake remains found among the 68-million-year-old Mahajanga fossils would seem to argue for an early origin. Yet their morphology was quite different from Madagascar's living boas, and some molecular estimates put the origin of the living species long after Madagascar became an island. The jury is still out, and one of the questions running through my mind those many years ago still awaits a definitive answer.

Turning to mammals, around 26 orders of living land mammals are recognised in the world and modern Madagascar is home to endemic species belonging to just five. That means there are plenty of gaps.

First to arrive were lemurs, members of the order Primates. The lemur ancestor came ashore between 60 and 50 million years ago. Aye-ayes (*Daubentonia madagascariensis*) most likely split off early from other lemurs, although it has been suggested that they arrived separately. Either way, with plenty of time to diversify it is curious that there is only one living aye-aye and one larger, recently extinct species known only from sub-fossil remains. Perhaps there were more in the past, and they disappeared without trace. All lemurs today live in forests, and feed on some combination of leaves, flowers and fruit, or insects and gum. It sounds quite dull, but they have unusual and interesting specialisations – like the ability of three species (*Hapalemur spp.*)

79

Mammalian crossings of the Mozambique Channel
(drawing by Luci Betti-Nash)

to feed on cyanide-laced bamboo. Curiously, few species are fruit specialists compared to primates in other forests of the world, perhaps a consequence of the erratic production or characteristically low protein content of fruit in Madagascar.

The behavioural, morphological and demographic oddities of lemurs, 'the lemur syndrome', have fascinated evolutionary biologists for years and are what first drew me to Madagascar. Females are about the same size as males (in most primates they are smaller) and socially dominant (the reverse is true in most primates), and social groups contain approximately equal numbers of each sex (females outnumber males in most primates). One could conclude that, not having read the textbooks, lemurs do not know the rules. But this is not an altogether satisfactory conclusion, and much effort goes into the search for other explanations. Less noted, the diurnal lemurs – species active by day – exhibit a remarkable array of colours. Beauty surely counts as

80

part of the lemur syndrome too, from the silver, honey and white of *simpona* (*Propithecus diadema*) and white, chocolate and burgundy red of *sifaka* (*P. coquereli*) to the blue-eyed gaze of a jet-black male and chestnut female *akomba* (*Eulemur flavifrons*) and the gracefully arching tail of a *maki* (*Lemur catta*), ringed with black and white stripes. What evolutionary pressures drove this kaleidoscope of colours? Another rabbit hole to disappear down . . .

Many scientists now recognise over 100 living lemur species. Twenty years ago, the number was 35. New molecular distinctions between populations previously considered to belong to a single species have driven this huge increase, and some question the validity of about half the new species. Unlike many of their diurnal relatives, the newly recognised nocturnal species are difficult to distinguish by eye. We humans – and the diurnal lemurs – perceive the world much differently from animals active at night. Colour means a lot to us in our daylight world whereas sounds and smells count for more at night, and the grey/brown colouring of nocturnal species may also help them escape the attention of owls and other predators. Firmly establishing their separate status as species on the basis of morphology and behaviour as well as molecular biology requires diligent effort.

Sub-fossil remains are the only evidence of 17 lemur species. Extinctions over the last few thousand years eliminated several species from five families that still have living representatives, and every member of three – Palaeopropithecidae (sloth lemurs), Archaeolemuridae (monkey lemurs) and Megaladapidae (koala lemurs). They all shared a common ancestor deep in the past with living lemurs and were in many ways like them, but they also expand our picture of lemur diversity. For one thing, they were bigger than any living lemur. The latter range from 30 g – probably the smallest primate in the world – to almost 10 kg, but if recently extinct species are included, the size range equals that of the entire array of primates in Africa and Asia – apes, monkeys and lorises – and far surpasses the monkeys of South America. Lemurs today are agile leapers who spend most or all of their time in trees and need forest to survive. Most of the extinct forms were adapted for life in the trees too, but with their much

Selected recently extinct lemur species compared to the largest living lemur,
Indri indri *(drawing by Stephen D. Nash, in Mittermeier 2010)*

larger bodies they would have clambered around much more slowly and likely came down to the ground instead of leaping from one tree to another, and monkey lemurs, in particular, probably spent more time on the ground than any living species.

After lemurs came tenrecs, members of the order Afrosoricida. Their common ancestor arrived between 42 and 25 million years ago. Tenrecs are less famous than lemurs and are represented by only one endemic family and a mere 39 species (though with more in the wings, awaiting formal names). They are all quite small, moreover, the largest weighing less than 2 kg and the smallest less than 3 g. But whatever tenrecs lack in fame, species numbers, or size, they make up for in diversity and oddity.

All tenrecs have very low average body temperatures and metabolisms – occasionally so low that an individual can be mistaken for dead – which fluctuate with outside temperature, almost like little reptiles. Some are underground digging machines, others scamper around in bushes and trees. Four species use their long tails as a fifth limb, dangling upside down by the tip wrapped around a branch. Members of another species swim along streams with webbed feet, their tail a rudder. There are tenrecs resembling hedgehogs that can roll up into an impregnable spiny ball, and tenrecs with spines tipped with small fishhooks that catch in the mouth or muzzle of would-be predators. They all use simple echolocation, and some use ultrasonic sounds produced by a special spine-vibrating organ on their back to communicate with one another. And some species reproduce at a rate unsurpassed by any other mammal. All these 'tenrec facts' are about animals that are hard to study and still not very well known as a result. One can only imagine the tenrec wonders still awaiting discovery.

The next ancestor of Madagascar's endemic mammals to arrive, between 26 and 19 million years ago, was from the order Carnivora. There may be as many as ten species today, all probably belonging to a single family, although further research is needed to confirm this. They range from small, agile, mongoose-like animals with low-slung bodies and short legs to the 10 kg *fosa* (*Cryptoprocta ferox*), which looks only like itself – a kind of short-legged mountain lion. (The common name of this species in English is fossa, but I use the Malagasy word.) *Fosa* are powerful predators, and a bigger version of this species (*C. spelea*), which preyed on some of the now-extinct lemurs, went extinct itself in the last few thousand years. In 2012, encountering a *fosa* lounging in my path on the way to breakfast at Kirindy in the western forests and mindful that they treat *sifaka* more or less as popcorn, I was glad not to be coming up against its extinct relative. Even so, I did not try to shoo it out of my way.

The ancestor of Madagascar's endemic sub-family of rodents, in the order Rodentia, arrived during roughly the same time span as the Carnivora ancestor. For me, the most intriguing of the 28 endemic rodent species living today is *vositse* (*Hypogeomys antimena*), known in English as the giant jumping rat even though it rarely jumps and is scarcely a giant. Normally it moves on all fours, breaking into a hop only when scared. It is certainly the biggest of the Malagasy rodents, about the size of a rabbit and with the look of one too, and there are sub-fossils of an extinct species that was larger still.

Hippopotamuses, from the order Artiodactyla (which also includes species like giraffes, camels and pigs), were once common in Madagascar and may only have disappeared completely within the last century. There were probably three species, one descended from the African common hippo and two from the pygmy hippo. 'Pygmy' is a relative term, however. Although African pygmy hippos are about half the height of common hippos and weigh much less, you still wouldn't want to mess with one on a dark night – and Madagascar's pygmy hippos seem to have been a bit bigger than their African ancestor. All three species were relative newcomers. Hippos cannot swim, though it certainly looks as though they can when dancing along the bed of a lake or river, and they do not float. Like that brave little African frog, it seems that their ancestors improbably rafted across the Mozambique Channel in just the last few million years.

One more order completes the roll call of land mammals. There was once a group of Malagasy mammals so oddly different from any living mammal they have an order all to themselves: Bibymalagasia. The name comes from the Malagasy word *biby*, meaning animal. The order contains two species (*Plesiorycteropus spp.*), known from sub-fossil sites along the island's western flank and in the central highlands. The remains are a few thousand years old, though palaeontologists agree that they are from an ancient mammalian group. They were very strange animals indeed, so strange that at one point it was suggested they did not exist at all and were, rather, a 'palaeontological chimera' – a collection of bones from different species. Today, debate centres on whether they were closely or distantly related to African aardvarks, evolved independently and merely looked like aardvarks, were actually giant tenrecs or were possibly related to both aardvarks *and* tenrecs. Were they part of the ancient Gondwana fauna or did they reach Madagascar by sea? The origins of Bibymalagasia remain a mystery for now.

*

If you have wings, flying is the easiest way to cross water. Indeed, you can come and go readily. This makes the evolutionary history of bats and birds at least as difficult to trace as that of accomplished rafters like lizards.

Fossil and molecular data place the origin of the world's major bat lineages about 50 million years ago. Bats presumably began settling in Madagascar after that, though little is known about the timing, and some arrived in the last few million years or the very recent past. Most or all arrivals flew across the Mozambique Channel from Africa.

Forty-six bat species are currently recognised, and the number keeps growing as field research continues. Over three-quarters of the species are endemic. Madagascar's fruit bats, all endemic, are a dramatic sight, whether soaring above the treetops by day, hanging quietly from the branches of their roost, or screeching and jostling like orange Christmas tree ornaments come to life. In contrast, the small, insect-eating bats are nowhere to be seen by day and move swiftly and silently at dusk. The most noise I ever heard from them was one summer at Bezà Mahafaly, when every evening hundreds of little bats made their way to the exit from their roost under the roof of a building in the camp. The sound of their claws clacking along under the corrugated metal was like that of heavy rainfall. Less poetically, the rooms below were thick with guano.

All but one of the lineages to which Malagasy bats belong have representatives around the world. This is not surprising for mammals that fly. But Madagascar stands out compared to other big islands. The diversity of its bat species is much lower and the proportion of endemic species much higher than in Borneo or Australia, taking account of those countries' far bigger surface area. This combination of low diversity and high endemism is puzzling.

Turning to birds, the puzzle persists. Madagascar has around 280 bird species, excluding marine forms, and well over two-thirds of them breed regularly on the island. (Only these are included in the table of living families and species on page 69.) Most species belong to just a few of the 63 families present.

The arrival of bird ancestors was evidently a long-drawn-out affair, with estimated dates scattered over the last 65 million years, although the majority seem to have arrived over the last four million years or so. Like bats, bird movements are not constrained by the direction of ocean currents.

The puzzle is that, like bats, the diversity of bird lineages is quite low compared to other lands of similar size, and seems to have involved relatively fewer colonisation events. Absent are several dominant groups in Africa such as hornbills, honey-guides and woodpeckers. Sub-fossil remains of some 20 extinct species have been found, including three mighty eagles, and others may have disappeared without a trace. Low diversity could be a result of recent extinctions or the workings of chance. But there may be more to it than that, as we shall see in the next chapter.

Madagascar's elephant birds (*Aepyornis spp.*) are the stuff of legend, the supposed model for the *roc* encountered by Sinbad the Sailor in *One Thousand and One Nights*. Sub-fossils, the occasional intact egg, and a lot of eggshell are what remain today to breathe life into legend. Four species are now recognised. Among them was the largest bird ever known, somewhere between 3 and 4 metres tall and weighing 275–400 kg, with eggs a massive 150–170 times the volume of a chicken egg. Evolutionary biologists long thought that the common ancestor of elephant birds and large-bodied flightless birds in Africa, South America, Australia and New Zealand roamed Gondwana on foot, too big to fly, and lived in Madagascar before it became an island.

This turns out not to be so. Molecular evidence shows that the closest living relatives of elephant birds are New Zealand's small, nocturnal kiwis. This is odd on the face of it – Madagascar and New Zealand are far apart geographically today and were

Mounted elephant bird skeletons. Left to right: Mullerornis modestus, Vorombe titans, Aepyornis hildebrandti. *On the far right, for comparison, the African ostrich* Struthio camelus *(adapted from Lamberton 1934)*

never very close to one another in the past. But it is now clear that about 50 million years ago elephant birds and kiwis shared a common ancestor small enough to fly – and fly a long way. Only after reaching Madagascar did these birds evolve to become larger and earthbound.

*

The evolutionary history of all these animals is inextricably and sometimes vividly intertwined with their lives today. My mind goes back to a morning almost 50 years ago. It was cold, close to freezing, at dawn in the southern forest. Chilled to the bone as I peered up at a cluster of white bundles in the trees, it was small consolation to know that by noon the temperature would soar above 90° F. On this morning, my goal was to accustom a new group of *sifaka* to my presence. When the sun finally lit up the treetops, the bundles resolved themselves into six sleepy animals. A yawn here, a half-hearted scratch there, and then with a few great bounds they moved up out of the shadows and into the sunshine. Local legend has it that *sifaka* worship the sun. It is easy to see why, for they will sit for an hour or more on cold mornings, facing the sun with their heads thrown back, arms flung wide, legs splayed and eyes closed. When the sun goes behind a cloud, as one they hunch over and turn back into so many white bundles. With scattered clouds in the sky, they really do look as if they are alternately saluting the sun and bowing down before it.

I shifted my position gingerly, and six pairs of eyes looked down straight at me. I froze. Nothing happened. A half hour passed and still nothing happened. The sun slowly warmed me as it rose in the sky, and suddenly all six animals leapt into action, bouncing from limb to limb as they hurled alarm calls at me. It was a brief onslaught. Away they went, having made their point, sailing through the treetops as I scrambled through the undergrowth trying to keep them in sight. Another day of habituating *sifakas* was underway, and in the back of my mind sat a question: why did they wait so long to flee? There is still a lot to learn about *sifaka* physiology, but we do now know that their energy turnover is very low indeed for a mammal their size. Was the half-hour pause between first sighting me and sending alarmed insults my way the physiological relict of an earlier time when the ancestors of all lemurs snoozed their way across the Mozambique Channel? I think so.

The past weaves its way into the present, and the present does not stand still. The parade of animals goes on, largely orchestrated by our own species for the last few thousand years. People have transported many new species to Madagascar, on purpose or by accident. Their introduction has brought new diseases and forms of competition to endemic species, threatening their survival and perhaps contributing to recent extinctions. I return to this in a later chapter, after people enter the picture.

The history of Madagascar's wildlife is a saga of change and innovation, from cannibal dinosaurs to paddling tenrecs. It is also a saga of connection and isolation. When reptiles dominated the world, they dominated Madagascar too. When an asteroid devastated the world, Madagascar was not immune. As mammals gained their ascendant place in the world, so they did in Madagascar. Contrasting with this connectedness, Madagascar's wildlife also had a character of its own from the very earliest days. Today, it is overwhelmingly the outcome of chance arrivals. Some animals did not survive, but those that did evolved into an extraordinary array of forms and lifestyles. It is as if the survivors improvised as they went along. Let us take a look at when, how and why that improvisation happened.

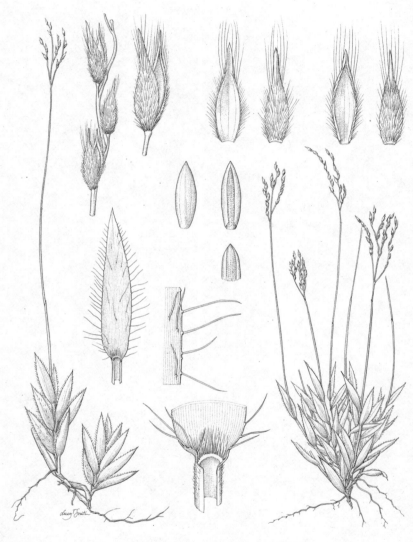

One of many endemic grass species, Yvesia madagascariensis
(drawing by Luci T. Smith)

CHAPTER 5

A Crucible for Evolution

Bedraggled primates scrambling up a beach to reach the trees, anxious lizards scurrying across sand toward the safety of undergrowth, exhausted birds plopping down in treetops . . . Reaching Madagascar was not for the faint-hearted. Looking back at the vicissitudes of life there over the 65 million years that followed, however, my admiration for the colonisers is equalled by my respect for their descendants.

Molecular biology transported us to a few animals making landfall far in the past, but what drove the evolution of a handful of colonisers into the rich diversity of species we see today, and when did it happen? Changes in the island's vegetation are surely part of the answer. Some were long and slow. Plants adapted to harsh droughts gradually replaced moist forests and prevailed for millions of years, only to recede with the return of wetter times; different types of vegetation slowly came to dominate landscapes as the modern diversity of climates emerged. Other changes happened fast, with forest, woodland and grassland boundaries shifting back and forth in the space of centuries or even decades in response to locally fluctuating conditions. Mapping the tempo and timing of wildlife evolution on to these events is still at an early stage, but the outlines of a map are beginning to appear.

The history of grasslands has an important place in this chapter, correcting the widespread and mistaken idea that there were none in Madagascar before people arrived. In the autumn of 2014, I had a firsthand glimpse of the research involved. It was a time of political upheaval and lawlessness in Madagascar and *dahalo*, bandits, were a much-feared presence in the countryside. Our car rolled to a halt in the shade of a tamarind tree an hour south of Andranovory in the southwest. The potholed dirt road, National Route 10, receded north and south in a meandering red ribbon. To either side, grasslands with a scattering of scrubby trees and shrubs stretched to the horizon. At this time of year, the height of the dry season, grasses are old and dry. Soon, cattle herders will set fire to them, encouraging a surge of new growth when the first rains fall. In a flash, the landscape turns intensely green and cattle fatten on the carpet of juicy young grass blades. But the moment passes quickly and, like this trip, my journeys are usually in the dry season. For me, these lands are gold, brown and yellow – the colours of old grasses – or seething red and orange, as fire licks across them.

Our driver Eugene Ramaroson, known as Zeze, paced anxiously. His task was to shepherd his charges – a band of Malagasy and foreign scientists – safely to Bezà Mahafaly, still four grinding hours away. He was not meeting with success. Several of his passengers were happily tramping further and further away from the car. With deep knowledge of the grasslands of southern Africa, William Bond had long been keen to see something of the open country of Madagascar. His first visit in 2006, with John Silander and his former students Joelisoa

Ratsirarson and Jeannin Ranaivonasy, produced a ground-breaking article arguing for the antiquity of grasslands in Madagascar. Now they were all in Madagascar together again, with me tagging along this time. And what a team they were; a towering South African, a towering American, and two more modestly sized Malagasy colleagues who between them fielded formidable expertise in grasslands, plant ecology and remote sensing (using satellites to scan Earth's surface). They were on a mission, and *dahalo* were not about to stop them. I followed William closely as he squatted down every few steps with exclamations of delight. What he saw was not just one or two brave species capable of withstanding fire but a multitude, several endemic to Madagascar – evidence for the antiquity of the island's grasses, and for natural fires not set by people.

Ecologists, botanists, palaeobotanists and palaeoclimatologists have uncovered many clues to the island's ancient vegetation. Their detective work employs several methods. One uses the remains of plants and other materials preserved in sediment cores. These cores capture a level of detail unequalled by other sources, but they span only the last 150,000 years at best and come mostly from the central highlands. Another approach depends on the molecular analysis of living plant species to establish their evolutionary relationships and times of origin. This is an exercise in reverse engineering, already encountered with animal species.

A different form of reverse engineering exploits links between vegetation and climate. Ancient plant (and animal) remains are widely used to infer past climates. In Madagascar, by contrast, the absence of fossils means that inferences operate mostly in the other direction. It works like this: the climate during a particular period is modelled from the broadly understood physical mechanisms controlling modern climates; the resulting reconstruction is matched to a region with a similar climate today, and the vegetation present under modern conditions is assumed to be what grew under those conditions in Madagascar in the past. The exercise involves many uncertainties, not least because climate is by no means the only determinant of vegetation, but triangulating between the various methods takes us as close to the history of vegetation as we can come for now.

*

My first serious encounter with Madagascar's plants came shortly after I arrived in March 1970. A day's drive brought me to a forest in the northwest that I had chosen as one of two places – the other in the south – I would spend the next 18 months. Jean Albert, then head of the Ampijoroa Forest Station, showed me around early the next morning. He was knowledgeable and kind, identifying tree species for me in the canopy over our heads as we walked trails together in the green and verdant forest.

After that, I spent the afternoon watching lemurs munch on leaves that all looked the same to me, despite Jean Albert's introduction and no matter how hard I tried to tell them apart. Feeding behaviour was central to my research on how conditions in the lush northwest and arid south shaped the lives of *sifaka*. What if I never succeeded in telling one leaf from another? That night, only the prospective humiliation of hightailing it back to England kept me in my camp bed under a big mosquito net. Today, I know those trees at Ampijoroa are among at least 14,000 plant species or candidate species in Madagascar, close to 90 per cent of which occur nowhere else in the world. And I can distinguish between at least some of them. Then, they were a daunting bewilderment.

Thank goodness for botanists. The best overview of Madagascar's vegetation today is *The Atlas of the Vegetation of*

Madagascar. It is the product of a mighty collaborative effort involving the analysis and synthesis of satellite images, and the collection and collation of information from the field. Most of the work was done between 2003 and 2006. Although I call it the island's vegetation 'today', the sad truth is that changes over the last decade or so mean that it is already really a picture of 'yesterday'.

The maps combine sophisticated technology, botanical expertise, and a huge amount of work. Data from a NASA satellite provided measures of greenness and the temperature and reflectivity of the surface; from these came estimates of canopy cover and leaf area, generating a broad-strokes picture of land cover over the entire country. Higher-resolution images were used to strengthen initial results, their accuracy checked during gruelling field trips.

With delightful attention to detail, the Atlas's authors were at pains to explain their choice of colours for the maps, which juggle competing demands between readability, printing considerations and what the gazing eye perceives. The resulting trade-off means that although the eastern rainforest is a pleasing shade of green, one has to imagine the deep green of the coastal mangroves (plum-coloured in the Atlas), southern forest silvers (olive drab or strident yellow), and golds and greens of the southern grasslands (pink and mauve). This is not to quibble with the authors' choices. It is, rather, to make a point: the range of vegetation on the island far surpasses the number of distinguishable shades of silver, gold and green available for illustration. No wonder they resorted to plum and pink!

Transitions in climate and vegetation between one region and another in Madagascar are gradual, except when they are not. Atop a pass in the far southeast corner of the island, just ten footsteps took me from the warm, wet conditions of the east to the island's most arid conditions of all. It happened during a trek through the forests of Andohahela National Park with colleagues in 1989. It was astounding.

Our walk began east of the island's long spine, at the small village of Isaka Ivondro. We climbed up and up, always westward, for much of the day. It was warm and humid, so humid that a universal dampness enveloped us all after a few hours. Birds flitted through the trees, though the lemurs we had hoped to see kept their distance. A host of hungry leeches was by far the most abundant wildlife. A single unfed leech is small and thin. By the thousand, they looked like a waving lawn on the sodden leaf litter of our path. It was a voracious lawn, and boots, trousers and leech-gaiters proved no impediment. Mark Pidgeon, the expedition's bird expert, pulled 80 leeches off one leg and then stopped counting during a mid-morning break. The stream we sat beside turned red with blood. Meanwhile, our Malagasy colleagues were leech-free, having worn shorts and smeared their legs with a green paste of specially selected leaves before we set out. That night we slept on the sandy floor of a rock shelter. It was low and narrow, an unprepossessing campsite at first glance. But once the fire was lit and a pot of rice simmering away, it felt warm and cosy. Rain fell in buckets all evening. A million leeches waved along the shelter's edge – but none crossed the drip line. Contemplating how long they must wait for a passerby, and a meal, I was left with a reluctant admiration for these resilient little creatures, able to go without food for several months at a time. I drifted off to sleep thinking that this was a place my archaeologist husband should excavate someday. Surely people had slept there long before us.

The second day was a long slow climb and then a steep one up to the pass. The only sounds in the forest were the thud of our footfalls on the damp litter and occasional birdcalls. Suddenly, about noon, light and sunshine lit up the path ahead. Thud went our feet, and sweat dripped off my chin. Then a crackle mixed with the thud, and I felt a whisper of dry wind. Ten more footsteps and we reached the top of the pass. The litter crackled noisily, a warm dry wind blew away sweat, and westward, down from the pass, a sunlit vista of silvery dry forest opened up. The climate and rainforest of the east had vanished behind us.

Where and when did the plants originate that compose today's grand diversity of vegetation? Almost a century ago, Henri Perrier de la Bâthie pointed out how closely many of Madagascar's plants resemble species found elsewhere and, based on these resemblances, he assigned them to groups according to their likely land of origin. These origins, he proposed, included not only nearby Africa and islands in the West Indian Ocean but also places much further away. Plants do not walk but, if he was right, they certainly move around a lot. The evidence was limited, however, and the matter remained one of intense speculation.

With acceptance of continental drift as a feature of Earth history, two possible explanations emerged for similarities between plants on the island and in distant regions. One was that Madagascar was a kind of crossroads in Gondwana that incorporated species from all the lands around it. The other was that plants mostly made their way to Madagascar by sea or air after it became an island. But there was no way of knowing which or what combination of these scenarios best represented what actually happened.

Molecular and morphological approaches used today in tandem show that Perrier de la Bâthie's conclusions, reached from morphology alone, turn out to be largely correct. The closest relatives of many endemic plant lineages in Madagascar are in Africa. Some have strong connections with Southeast Asia or India, and a few have even more distant geographical links to Australia and South America. Not surprisingly, the closest relationships of all are with plants nearby, on the little islands of Mauritius, Reunion and Rodriguez just east of Madagascar, and the Comoro Islands to the north.

Molecular evidence has also begun to provide estimates of when species or whole lineages diverged from one another and, thus, how they arrived. The age of the last common ancestor shared by modern Malagasy plants and their relatives in other lands provides a crucial clue. As for animals so for plants – if the common ancestor lived *after* Madagascar became an island, it must have journeyed to (or from) the island across water. If it lived *before* Madagascar became an island, then it likely originated there or made its way there overland when Madagascar was still part of Gondwana.

Most tree lineages on the island today originated no earlier than 40–45 million years ago, and many more recently than that. The ancestors of some would have travelled in mats of vegetation; the seeds of others were blown by the wind, or transported by birds; fruit such as coconuts may have bobbed their way across wide stretches of ocean, carried along by currents. A few lineages predate the asteroid impact 66 million years ago, but only one has a founding ancestor old enough to predate the break-up of Gondwana. Far from being of great antiquity, Madagascar's modern vegetation is composed of quite young plants in evolutionary time that arrived from near and far.

Take *Canarium*, for example. Trees belonging to this genus grow in many tropical and subtropical forests around the world. Madagascar is home to between 6 and 12 endemic species (the taxonomy is still being worked out) and they are a dominant presence in many moist forests, making up a third of the volume of trees in some. Several lemur species relish *Canarium* fruit, and for some it is a major food. Diverse, dominant, and a key species in some lemur diets, *Canarium* would seem a good candidate to be an ancient denizen of the island. Yet the entire array diversified from a colonising ancestor only 8 or 9 million years ago, probably transported from Southeast Asia by birds or ocean currents. Lest this seem an improbably impressive journey, it is worth noting that pumice stone thrown up by the 1883 eruption of the Krakatoa volcano 6,500 km away in Indonesia washed ashore on the east coast of Madagascar six months later. If pumice stone could make the trip, why not *Canarium* seeds too?

*

When Madagascar began its arduous passage through the arid belt around 65 million years ago, drought-adapted plants would have gradually replaced moisture-loving species and one can imagine the island's vegetation looking increasingly like the spiny forests of the south today. A scattering of more hospitable forests must still have persisted in sheltered pockets of land, however, providing refuge for the ancestors of today's frogs and other recently arrived creatures unable to tolerate harsh conditions. As the island emerged north of the belt about 30 million years ago, the moderating climate would have sent drought-adapted plants into retreat, and the far south became their last refuge.

The Didiereaceae family dominates spiny forests in the south today. It includes the silvery trees studded with spines that first set me on my path to Madagascar. Members of this family thrive under arid conditions. Molecular evidence suggests that it originated between 56 and 34 million years ago, and its ancient members probably spread across much of the island inside the arid belt. The Didiereaceae retreated south as conditions became wetter, and mostly died out. One species survived, it seems, diversified anew 10–15 million years ago, and gave rise to the 11 species alive today. It is too bad. I wish the forest that captured my imagination fifty years ago in a Cambridge classroom was a direct vestige of the most ancient forest on the island. Unfortunately, the evidence points to a more complicated pathway to the present.

Alluaudia procera, *a member of the Didiereaceae family, in the southwest (photograph by Ed Lowther)*

About 7 million years ago, with climate conditions drying globally, vast landscapes of open country began to appear in the tropics and subtropics. What happened in Madagascar? What you ask has a strong bearing on what you look for, and what you are likely to find. Until recently, no one was asking if Madagascar had ancient grasslands and so no one was looking for evidence of them. A few decades back, a distinguished botanist conceded that climatic events could perhaps be 'blamed' for the origin of some of Madagascar's grasslands. Perhaps they could. It is an odd verb though, and happily the era in which grasslands are automatically someone's or something's fault is receding.

Well-studied grasslands on other continents provide a roadmap for research in Madagascar. They come in many forms and are called by several names – savanna in eastern and southern Africa, prairie and plains in North America, pampas in South America, and steppe in Eurasia. Grasslands is probably the term most commonly used today to encompass habitats where the grass is taller than a person or under a centimetre high, places where a sea of grass sweeps to the horizon with not a single tree in sight, and places with quite a lot of trees and shrubs. When does vegetation stop being grassland and start being woodland or forest? There is no clear answer, reflecting the perennial challenge of imposing discrete categories on an endlessly varying world. For now, it is a challenge we can duck. In Madagascar, the main question is whether vegetation existed that could reasonably be called grassland of *any* description.

Climate and soil have a strong influence on what kinds of plants will flourish in a particular region, yet models predicting global vegetation based on these factors alone indicate that far more of Earth's land surface should be covered by forest than is actually the case. This is partly because people have been clearing forests for thousands of years but also because the models ignore the roles played by other forces. Climate, soil, fire and herbivores are often all at work together on a grassland, moulding its extent and character as they fluctuate and interact with one another. This can bring about major shifts over the course of a few centuries, or even decades. In Kruger and Limpopo National Parks in southern Africa, for example, minor changes in climate and fire regime have shifted the landscape back and forth several times between grasses, trees and shrubs over the past few centuries. At Amboseli in East Africa, grasslands liberally scattered with trees became virtually treeless in just a few decades. It was not that the growing population of elephants ate all the trees – they simply bulldozed them.

Grasslands everywhere were long the underdogs of vegetation, viewed as temporary formations on their way to becoming forests, but dynamics like those described in eastern and southern Africa pushed back on the idea of open country as a kind of accident of nature that will naturally right itself. Ecologists developed this proposition, called succession theory, in the early twentieth century and it held wide sway for many decades. One has only to watch abandoned fields in New England successively colonised by weeds, shrubs and trees over the years to appreciate its explanatory power. But the power has limits. Open-country vegetation does not always march inexorably toward becoming forest.

Evidence for the shifting dynamic between grassland and forest comes from pollen grains and traces of photosynthesis left behind by plants and preserved in ancient soils. Photosynthesis refers to the way plants use the sun's energy to make sugars out of carbon dioxide captured from the air by their leaves and water captured by their roots. It takes more knowledge of chemistry than I possess to understand fully the differences in how they accomplish this, but the basic outline is straightforward and helps make sense of the labels given to the two most common processes. These processes are called C3 and C4 for short, and the terms are commonly attached to the plants themselves. The C3 pathway yields a sugary molecule containing three carbon atoms, and the C4 pathway four carbon atoms. A third pathway, Crassulacean Acid Metabolism (CAM), is mainly found in succulent plants, which are able to switch between C3 and C4 pathways.

C3 is the ancient pathway of terrestrial plants. About 85 per cent of living plant species are C3, including almost all trees and shrubs as well as some grasses. C4 plants evolved more recently; most are grasses, and they make up nearly half the grass species in the world. C4 grasses do well under dry, hot conditions, and C3 species in wetter, cooler environments. Plants leave plentiful geochemical traces of these pathways in soils of all ages, in the form of two distinctive, stable carbon isotopes. A high proportion of the carbon isotopic signature of C3 plants is interpreted as indicating wooded habitat or forest, whereas a high proportion of the C4 signature indicates open country. This interpretation may not always be correct. An isotopic signal showing a mixture of C3 and C4 species could mean tropical grassland dotted with trees and shrubs, cool temperate grassland with a mix of C3 and C4 grasses, or even tropical grassland with C3 and C4 grasses (though this last mixture is not in fact found today). Still, carbon isotope ratios are often the only evidence available and, caveats notwithstanding, they are widely used as a result.

The conditions enabling C3 and C4 grasses to establish their first footholds were evidently different from those triggering their expansion. Although each expansion occurred in a single broad wave on a geological time scale, ecologically it was far from simultaneous. The first C3 grasses appeared on several continents about 65 million years ago, living inconspicuously on forest floors for many million years before spreading to form grass-dominated habitats in some regions. C4 grasses are younger, some 30 million years old, and began expanding around the world only about 7 million years ago. They spread further and faster than C3 grasses ever did, however, and vast C4-dominated grasslands opened up in the tropics and subtropics. Their extent waxed and waned under shifting global conditions, but the great grasslands that delight safari-goers and fascinate ecologists today are descendants of these ancient landscapes.

Continents separated by wide oceans share many of the same C4 genera, and even the same species. Oceans do not constitute much of a barrier to grasses, and suitable conditions are a more important influence on their spread than distance of travel. Winds, birds, and rafts of vegetation must have done an excellent transport job during the expansion of C4 grasslands, and it should not have been much of a challenge for grass seeds to reach Madagascar. But did they?

In their article arguing for the antiquity of Madagascar's grasslands, William Bond and his fellow plant ecologists – the source of Zeze's anxiety on the road to Bezà Mahafaly – compared the number of C4 grass genera and endemics in Madagascar with those in southern Africa. The numbers should be much lower in Madagascar if the spread of grasslands was recent, but their findings did not fit this scenario. In fact, the diversity and ecology of Madagascar's grasses closely matched those in southern Africa. Aware that this flew strongly in the face of received wisdom – and, indeed, there was a lot of push-back when the article appeared – they concluded: 'We suggest that biologists should take a fresh look at Madagascan grass-lands'. A growing number of botanists and ecologists have begun to do just that.

One of them is Maria Vorontsova. I met her for the first time at the Royal Botanic Gardens, Kew, in March 2017. Her name sounded Malagasy to me, and waiting in the reception area, I expected to meet a Malagasy researcher. When a fair-haired young woman arrived to greet me, I assumed that Maria must have an assistant. My mistake. Names of Russian derivation can sound a lot like Malagasy ones, I learned, and Maria's family roots are in Russia. She assured me later that Malagasy make the same mistake.

Maria is a self-declared grasses fanatic, a Malagasy grasses fanatic in particular. Together, we pored over sheets of metic-ulously pressed specimens as she showed me the distinctive features of each. My mind briefly wandered back to Jean Albert in 1970, pointing out differences between tree species I could not tell apart. These grasses too looked much the same to me. It is a good thing I did not become a botanist and Maria did. She and her colleagues expanded on the earlier research, and they have brought our knowledge of Madagascar's grasslands a long way in the last few years.

Far from being isolated from ancient grassy events in the rest of the world, Madagascar was very much a part of them. The diversity of Malagasy grass species is indeed high – 350 or so C4 species and 175 or so C3 species, and new ones continue to be discovered. A higher proportion of these species are endemic than in most other regions of the world moreover, and different parts of the island have their own distinctive grass communities. Grasses did not spread across Madagascar from a single area. They colonised the island repeatedly and long enough ago for local specialisations to evolve.

Cédrique Solofondranohatra collecting grasses
(photograph by Maria Vorontsova)

Just under a third of the C4 grass species identified so far
are endemics, compared to over two-thirds of the C3 species.
Does that mean C3 grasses arrived earlier and had more time
to evolve unique features? For the most part, this does not seem

to have been the case. Molecular clocks and well-dated plant fossils indicate that the ancestors of almost all modern endemic grasses reached Madagascar from Africa between 7 and 4 million years ago, when grasslands were expanding globally. C3 and C4 grasses colonised and diversified across the island's varied landscapes differently, though. A limited array of C3 grasses prospered in moist and shady forests, as in other regions of the world. Meanwhile, a host of opportunistic C4 ancestors took root in open conditions wherever they occurred, and their descendants are important constituents of treeless grasslands and open woodlands in Madagascar today.

There were grasslands in Madagascar millions of years before people arrived; of that there is no doubt, although there is still much to learn. Were natural fires and grazing by wildlife important in maintaining them, as in other regions of the world? If Madagascar's endemic grasses tolerate frequent fires and grazing, it suggests these pressures existed millions of years ago and grasses had long enough, evolutionarily speaking, to adapt to them. If not, it would lend support to the idea that people were responsible for burning landscapes for the first time and that their livestock created grazing pressures never before experienced. Certainly, there are many grasslands today composed of few species, most of them not endemic, in formerly forested areas. People make intensive use of them for pasture, burning them annually to encourage new growth. But there are others where the human hand is less heavy, grazing pressure is low, and the fire return interval is several years – much like African grasslands of undisputed antiquity, where fire is recognised as a natural part of the system.

How do Madagascar's endemic grasses fare under differing conditions today? Maria and her colleagues set out to answer this question. Selecting 60 grassy sites in five areas across the island, they tagged them as having low, medium or high levels of burning and grazing, and collected, identified and counted the grasses from each. If endemic grasses are well adapted to these pressures, you would expect to find them even in areas where the pressures are intense. If they are not, you would expect instead a few, sturdy, introduced species with a long history of exposure in some other region of the world. What they found was that the diversity of grass species was no lower at sites exposed to frequent fires, and grasses at all three types of site were derived from local grasslands. Endemic species did well in the presence of fire, pointing to a long history of exposure and adaptation. In fact, many endemic Malagasy plants have similar abilities.

Although grasses appeared not to do as well in the presence of intensive livestock activity, intriguing new evidence from the central highlands suggests that the tolerance of some endemic species for grazing is higher than initially concluded. In this study, the team sorted samples from 71 grassland sites categorised according to several criteria, including the frequency of fires and distance to the nearest road – a proxy for grazing pressure. Two distinct groups of grasses emerged from their samples, one with a grazing and the other with a fire 'profile'. Grasses in the first group form mats – grazing lawns! – composed of short plants with wide leaves that are full of nourishment. Fire-tolerant grasses, on the other hand, are taller and less dense. They burn easily and regenerate well after fires, but their tough, narrow leaves are unappetising fare. Half or more of the species in each group turned out to be endemics, and their profiles corresponded with surrounding conditions: grasses with a grazing profile flourished under intense grazing pressure, while fire-tolerant grasses grew in areas that burned frequently. In short, research thus far points firmly to the conclusion that natural fires were important in the evolution and maintenance of

Madagascar's grassland systems, and suggests that an ancient grazing community played a role too.

What did these grasslands actually look like – a treeless sea to the horizon, a landscape scattered with trees and shrubs, or a vista of grassy woodland? I consider this question further in the next chapter but for now the answer is probably all three, their character shifting with time and place. It will take more research to assemble a full palette of colours for the wide expanses currently allocated to 'degraded grasslands', coloured pink and mauve, in *The Atlas of Madagascar*, but enough is known already to be sure that the rainbow of colours will expand in future editions.

*

Taking a great leap forward, we move from the spread of grasslands several million years ago to a time close to our own. From about 2 million years ago, global fluctuations in temperature brought alternating warm and cold periods to the tropics and subtropics. In Madagascar, a 40-metre-long sediment core from the bed of Lake Tritrivakely captures their impact for the last 150,000 years or so, shorter cores from this and another lake nearby fill out the record for the last 11,000 years, and sub-fossil sites add yet another dimension.

Lake Tritrivakely, in the crater of an ancient volcano in the central highlands, provides the oldest detailed record of climate and vegetation for any site in Madagascar. The sediments accumulated during several global cycles of warming and cooling, and diatoms, pollen and charcoal preserved within them, translate these cycles into a record of events around the lake and their approximate timing. Diatoms are like weather gauges. These tiny algae live by the thousands in almost all habitats where there is water, from oceans and lakes to mosses and soils. Rainfall and habitat determine the particular species likely to occur, and the shells of dead diatoms are virtually indestructible. Lakebeds and swamp bottoms are commonly vast diatom graveyards as a result. The combination of sensitivity to environmental conditions with good preservation means that these anonymous little creatures are stars of environmental reconstruction. The Tritrivakely diatoms signal that droughts of varying severity enveloped the central highlands at roughly the same intervals as in Africa. From the perspective of deep time, these episodes were minor changes. From the perspective of the plants, animals and, eventually, people contending with them, it was quite another matter.

Pollen takes up the story. Matching pollen grains from a sediment core to a 'library' of modern grains makes it possible to identify the species that produced them. The size of the area sampled by a core cannot be known with precision – it depends how far the pollen grains within it have travelled, and pollen from different plant species is not equally likely to be 'captured'. But comparing and calibrating preserved pollen with modern pollen in modern environments helps reduce this uncertainty. Pollen grains from Tritrivakely reveal that vegetation in the central highlands shifted repeatedly between forest, a mosaic of grassland, woodland and forest, and heathland.

Palynologists, people who study ancient pollen, occupy a place in my pantheon of heroes. Identifying and counting microscopic pollen grains by the thousand is a task of unimaginable tedium to me. In fact, it is this lemur-watcher's idea of a nightmare. Then again, my husband Bob would remark that he could not imagine spending twelve hours a day watching animals do nothing. At least I could be sure they were there, I would retort, which was more than could be said for the fragments of pottery he spent his days searching for. Choosing a field of study is perhaps partly a matter of deciding on your preferred form of boredom, and I am grateful some choose to count pollen grains.

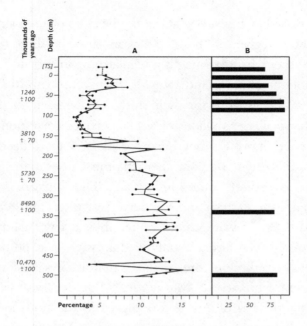

Fluxes in charcoal and grasses at Lake Tritrivakely over the past 10,000 years: percentage of charcoal per dry weight of sediment sample (A), and percentage of carbonised particles derived from grasses (B); (TS) is the top sediment above the core (adapted from Burney 1987c)

Analyses of charcoal come into play next. Ashes drifting through the air and charcoal from burned wood and leaves settle in lake-beds and peat bogs during and after fires. They form thick, dense layers when fires are frequent, and there these layers sit as sediments build up on top of them, awaiting a researcher armed with a tool for drilling cores. Charcoal 'spikes' examined in the Tritrivakely core spanning the last 11,000 years signal periods when fires blazed frequently long before there is any trace of people in the central highlands. Burned grasses and wood from trees are distinguishable from one another in charcoal, and it turns out that burned grass makes up much of the charcoal in spikes studied closely. Grassland was widespread and fire, the consumer, surely played a role in maintaining it. Indeed, spikes in a core drilled in a peat bog at the northern tip of the island were actually much higher around 32,000 years ago than those recorded at modern sites where burning is a regular human activity.

Our notions of time and rates of change are captives of the methods available to us for their study. Changes far in the past seem to happen over millions of years, perhaps corresponding to the real pace at which they occurred or perhaps reflecting the limits of our ability to pinpoint them in time. Closer to the present, more precise methods come within reach. The Tritrivakely record tells us that, as in eastern and southern Africa, change sometimes happened over a few thousand rather than a few million years – and possibly even faster, if only we had ways to capture it.

Analysis of sediment cores makes a major contribution to the environmental history of Madagascar, but the view from the window it opens is of just a few places in the not-too-distant past. Conditions for the accumulation of sediments are not as favourable in most regions of the island; fewer cores have been extracted, and they span less time. It is unlikely that fires periodically raged across every landscape. Fire does not take hold in the soggy litter of eastern rainforests, for example, and it finds little fuel on forest floors in the south. In the central highlands and the west, by contrast, the weather provided excellent conditions for seasonal fires, and powerful, dry-season lightning storms or volcanic activity lit the match. It is a reminder of Madagascar's wide variety of environments past and present, and emphasises that our understanding of the island as a crucible for evolution is still quite limited.

Sediment cores have one other drawback: the bones of big animals do not fit into them. Most sites where sub-fossils have been found are in the west and south of the island, and most are caves or sinkholes where bones tend to accumulate in a jumble, making it hard to sort out context or chronology. A newly discovered site at Tsaramody high up in the central high-lands is a rare exception. In a high-elevation natural basin walled in by ancient volcanoes, a sequence of deposits going back about 19,000 years preserves a record of both plants and animals.

Tsaramody was a wetland, with grassland, woodland and forest spreading and retreating over time like at Tritrivakely. Hippopotamuses, crocodiles, tortoises, elephant birds and water birds lived there. No trace of arboreal lemurs has been found, only the remains of *Archaeolemur*, the most terrestrial of the giant lemurs. At least during this period, the central highlands was likely a barrier to animals that depended on continuous forest for their living.

*

As a general matter, upheavals in geology and climate drive bursts of species diversification while stable conditions give rise to more gradual changes. Researchers have singled out two periods of abrupt change in Madagascar when the rapid splitting and emer-gence of new species would be likely. The first was when the island emerged from the arid belt about 30 million years ago. Moist forests spread as a balmier climate took hold, and perhaps this presented an opportunity for animal species to expand their ranges and diversify. The second period was much more recent, between 2 million and 12,000 years ago, a time of abrupt environmental shifts. Forests were widespread under warm conditions but retreated to small refuge areas when temperatures cooled. Under this scenario, animal populations isolated in separate refuges drifted apart genetically, diverged in behaviour and anatomy, and ended up as locally endemic species. When conditions warmed again and forests spread, the populations would have re-encountered one another as strangers – different species now.

Did ancient opportunity, recent 'moments' of abruptly changing conditions, or the steady tick-tick-tick of gradual evolutionary processes give rise to the diversity of wildlife we see today? Each offers a possible timeframe and mechanism whereby new species might arise. Isolation of one sort or another appears always to have been involved – from a river that could not be crossed to a fragment of forest turned into a prison by surrounding grasslands – but the molecular evidence suggests that no single geological or climatic event triggered evolutionary diversification. A one-size-fits-all approach does not work for the spiders, chameleons, geckos, skinks and lemurs studied in detail so far. Some analyses indicate that the process was continuous, with new species accumulating gradually over time, while others conclude that it happened early and slowed down as new ecological opportunities were exhausted.

One point seems clear. Temperature fluctuations and oscillating boundaries between forest and grassland over the last 2 million years were not important drivers of diversification – *except when they were*, pipe up the mouse lemurs. Mouse lemurs initially split into three groups 9 or 10 million years ago, but five of the 24 or so species alive today diverged from one another within the last half million years. Climate-driven changes in vegetation and the resulting isolation of populations from one another in the recent past seem the best explanation for this. Take *Microcebus lehilahytsara*. This mouse lemur is found along the edges of the eastern rainforest, and in forest fragments in the central highlands separated from one another by treeless grasslands. The scattered populations are genetically quite different from one another, well on the way to becoming two or more separate species. They make another point too. Animals from different forest fragments should still be pretty much alike genetically if the central highlands were continuously forested until people settled there in past centuries. *M. lehilahytsara* has been dubbed a 'telltale species' for the presence of grasslands before people arrived.

Evolutionary processes are still at work today, even without sharp distinctions between forest and grasslands. At Bezà Mahafaly, for example, some mouse lemurs are red, others brown, and still others grey. Emilienne Rasoazanabary, a young Malagasy scientist who studied these animals for several years, initially determined that there were three species in the forest. It was a reasonable conclusion. They did not simply look different, they were different sizes and their hands and feet were shaped differently. Yet genetically, it turns out, they are indistinguishable. What are we to make of this? One possibility is that there is only one species, unusually variable in coat colour and morphology. A second is that one species is in the process of replacing another with some hybridisation between the two, and a third is that the population is at an early stage of dividing into two or three separate species. More work is needed to figure out which, if any, of these possibilities explains the odd mish-mash of features.

Watching nocturnal lemurs is not my favourite activity. You do not see a lot. Finding an animal in the dark means walking through the forest and hoping to pick up its eyes, aglow like hot coals in the beam of a headlamp. Depending on the species, the animal either just sits there and watching it quickly (to me) gets quite boring, or else it leaps off and you lose it. Mouse lemurs are in the second category, and Emilienne is a person of far greater fortitude than I. Bezà's enigmatic mouse lemurs certainly lure me into the forest at night more often these days but, much more important, they are a counter-point to the idea that everything of significance happened long ago. Madagascar's stunning species richness in many lineages, including those of the lemurs, is the cumulative result of many events during its turbulent geological and climatic history. And the saga continues. You can witness speciation happening in Madagascar with your own eyes.

The aye-aye, Daubentonia madagascariensis
(drawing by J. Wolf, in Owen 1863)

CHAPTER 6

Familiar Tasks Done Differently

Madagascar is slowly yielding clues to how a small number of colonising animals gave rise to such diversity in the present, but this does not explain the distinctive or downright odd characteristics of many modern species. The uniqueness of Madagascar's wildlife can be over-egged, and its amphibians, reptiles, mammals and birds have a lot in common with their relatives in other regions of the world. And yet . . . Sixty-six million years ago, the young island's fauna was already like-but-not-quite-like that of Gondwana. Then came the asteroid hit, things started over, and Madagascar's modern fauna is even more distinctive than its ancient inhabitants were.

The life history patterns of many living species are among these unusual characteristics, and Madagascar's unpredictable climates help explain why. Life history patterns encompass tasks fundamental to a species' persistence: how long an individual spends in the uterus or egg, how fast it matures, how likely it is to die young, how long it lives if it reaches maturity, how often a female gives birth, and the number of young she produces at a time. The strange features of *sifaka* life cycles first caught my attention, but it now seems that odd life history patterns are widespread.

Many animals in Madagascar also make a living in unexpected ways. The sweepstakes nature of long-ago colonisation events is likely key to the evolution of their distinctive lifeways, for few land animals reached the post-asteroid island and nature improvised with what was at hand. 'Look for empty niches!' This was the late Alan Walker's exhortation when I stopped in Uganda to meet him on my way to Madagascar for the first time in 1970. Alan would go on to become a distinguished anatomist and palaeoanthropologist and, although we had shared a PhD supervisor, our real bond was a deep fascination with lemurs. He wanted me to find evidence that living lemurs had expanded into roles left vacant by the extinction of the largest-bodied species. I never did, though it might explain why brown lemurs feed heavily on leaves yet neither their teeth nor digestive system are well adapted to chew up or digest them. But Alan's exhortation stuck, and it was a short step to start contemplating the empty niches open to the first lemurs arriving in Madagascar and, from there, the possibilities for creatures of all kinds.

The idea that animals in Madagascar seized evolutionary opportunities to fill unoccupied niches has mostly centred on forest-living species, and the search for species living in open country inhabited by other kinds of animals elsewhere is new. Ungulates evolved to feed on African grasslands, kangaroos in Australia, and giant ground sloths in South America. We know now that grasslands go back many million years in Madagascar. Was there once a community of open-country animals there too? Scepticism about the received wisdom – that there were no ancient grasslands – kindled my initial interest in searching out species that might have undertaken grazing and browsing tasks. Learning more about their possible role in maintaining the grassland draws me to them today.

*

Sometime during the last 8–15 million years, Madagascar's climates became markedly unpredictable. For animals, unpredictable conditions can have brutally direct consequences – like drowning when a dry river bed suddenly becomes a roaring torrent, freezing to death in an unexpected cold snap, or overheating when temperatures soar. The effects can also be indirect, with animals starving to death when food sources unpredictably collapse. Either way, they present species with an evolutionary choice between putting all their eggs in one basket, or not: 'reproduce fast lest you die', or 'go slow and hedge your bets'. Natural selection generally favours speed when adults bear the brunt of unexpected bad conditions and slowness when their offspring do, pushing species to evolve extremely fast or extremely slow life histories. The lifeways and behaviour of many species in Madagascar seem less odd viewed from this perspective, and 'extreme' life history patterns are also common in other tropical lands with highly unpredictable rainfall.

On the fast side, a good place to start is with mouse lemurs. Females begin to reproduce when they are only 12 months old, and have litters of two or three infants twice a year when conditions are good. This makes them the 'fastest' primates in the world. But they are far outshone by a species of tenrec. *Tenrec ecaudatus* has the biggest known litters of any mammal in the world, in fact – up to 32 infants at a time – presumably explaining why females have as many as 29 teats. Many females breed just once, and die soon after. This is life in a very fast lane.

Mammals are not the only fast-living critters. A species of

chameleon found in the southwest, *Furcifer labordi*, is convincing proof of this. Hatchlings grow quickly once they emerge from their eggs when the rains begin in November, and they reach full size (about 10 cm from snout to tail base) and begin mating within a couple of months. Once females have laid their eggs, individuals of both sexes start losing weight, moving more slowly, and falling out of trees. It looks a lot like ageing, and by February or March the entire adult population is dead. For the next eight to nine months, incubating eggs are this chameleon's only presence until the cycle starts again the following rainy season. *F. labordi* spends more time as a developing embryo in the egg than outside it. Of the roughly 30,000 species of four-limbed vertebrates elsewhere in the world, almost none are known to live and die within a single year. Before getting too carried away, however, I note that in the same forest as these lightning-quick chameleons lives a closely related species whose members look perfectly healthy and robust at the end of the breeding season. Single explanations rarely explain everything in nature.

Madagascar also harbours some of the 'slowest' animals on Earth – including the *sifaka* at Bezà Mahafaly. Fully half the females have yet to give birth by the time they are 7 years old. Compare that to your average *sifaka*-sized cat – which may start reproducing as soon as 4 months after its own birth! Slow to get going, once started they continue to produce a single infant at intervals for many years. Some females reach the ripe old age of 30 years, an unusually long lifespan for animals their size. *Sifaka* hedge their bets, living their lives in the slow lane. In bad years when the rains fail, many infants die within the first few months but their mothers survive, to try again another year. Most or all of the large-bodied extinct lemurs may have been bet-hedgers too. As a general rule, rates of reproduction are lower and generation times longer in large-bodied species than small-bodied ones. Reproduction cannot be studied in extinct animals, obviously, but teeth can. Rates of dental development suggest that, like their living relatives, the extinct lemurs matured and reproduced more slowly than other mammals their size.

The list of 'slow' mammals alive today grows as data accumulate: the endemic carnivores give birth only once a year and most give birth to a single offspring, strikingly 'slower' than their relatives in Asia and Africa. Remember the giant jumping rat that is not a giant and does not jump: it lives in monogamous pairs and gives birth to a single offspring once a year, a decidedly un-rodent thing to do. And then there is the curious case of the black-and-white ruffed lemur community studied at Ranomafana National Park in the eastern rainforest over a six-year period. These animals, among the most frugivorous of living lemurs, have evolved a strategy for coping with unpredictability like no other. On the one hand, they mature fast and females start reproducing quickly, with the shortest gestation length, largest average litter size and richest milk of any lemur. On the other, they have unusually long life expectancies and females continue reproducing throughout their lives. Complicating matters further, the females in this study gave birth to litters of 2–3 infants just once in six years – all within the same two weeks in the same year. The offspring of females who cooperated over infant care were more likely to survive than those with mothers who did not. This pattern of boom-bust reproduction combined with cooperation may allow females to 'make up for lost time' during non-reproductive years. One thing is really clear: ruffed lemurs add a whole new dimension to the simple dichotomy between fast- and slow-lane life history strategies.

This has all been about land animals so far, but let us turn briefly to creatures that fly. Environmental unpredictability may prove to be an explanation for the puzzlingly small number of endemic bat and bird lineages on the island. The Mozambique Channel was not as powerful a first filter for them as it was for land animals, but unpredictable conditions may have been. If so, bats and birds should have unusual life history patterns too. This is a prediction for now, but it seems to me there must be a connection. The island's climate system has been in place for millions of years, and so many land species have extreme life history patterns that it seems unlikely bats and birds avoided its consequences.

*

The aye-aye has attracted much hyperbolic prose. About to see my first one in the wild, I refused to accept the notion that it was the most magical of lemurs and was determined to be unimpressed. In 1984, a coastal trading boat deposited Bob, our two young daughters, a friend and me on a sandy beach on Nosy Mangabe. Nosy Mangabe is a small island about 4 km from the northeast coast in the Bay of Antongil, but the mainland looked far away as the boat chugged off into the distance with a promise to return in a couple of days. Today, the island is a nature reserve and no one lives there permanently.

Nosy Mangabe sits right in the path of cyclones that periodically demolish its lush rainforest, but the dense vegetation seemed all too intact as we set off along a narrow, muddy path in search of aye-ayes. Aye-ayes are nocturnal, and I had been told that early evening was our best shot at seeing them. We did not have to walk far. Within a few minutes, in the beam of a flashlight I caught my first sight of a dark, low-slung, hairy creature on a wide branch above our heads, a fitting passenger for a witch's broomstick. It stared at us with amber eyes for a

The aye-aye's hand (drawing by J. Wolf, in Owen 1863)

while and then sauntered off along the branch. I stood still in the moonlight, realising that I had to take back all my brave words about it being just another lemur.

An aye-aye's body looks as if it is put together with spare parts from other animals, or simply invented for fun: a fox's bushy tail, a bat's big, naked ears, and a rodent's large, continuously-growing front teeth. But their hands are unlike anyone else's. They are worthy of Halloween. Each fingernail is claw-like, the third finger is long and skeletally thin, and the fourth even longer but stouter. Aye-ayes use the spindly middle finger to tap on branches and, supposedly, their sensitive ears to pick up the sound of a hollow cavity beneath the bark that might contain a cluster of squirming larvae. Yet animals living in captivity find cavities even when filled with gelatin or acoustical foam. How they accomplish this remains an aye-aye mystery waiting to be solved.

Back on Nosy Mangabe, an aye-aye probes to the bottom of a nest of larvae with its long fourth finger and scoops one out whole. Biting off the larva's outer covering, the finger's owner eats it like an ice-cream cone, licking any flow of juices down its hand to get the very last drop. Years after our excursion, I nervously held out a fistful of larvae to an aye-aye at the Duke Lemur Center. I expected to be scratched, if not gouged. The aye-aye reached out and a long finger disappeared between my bunched fingers. There was a whisper of a touch. I opened my hand. The larvae were gone. Aye-ayes have been called Madagascar's version of the woodpecker. The way they extract larvae is different, obviously, and there are in fact other birds playing woodpecker-like roles. Still, they come as close to woodpeckers as any primate, an instance where evolution has apparently produced a strange convergence.

Ayes-ayes are not the only species to have stepped at least partway into others' shoes. Many tree species depend on insects to pollinate their flowers, but some, like baobabs, have big flowers, that produce large quantities of nectar and rely on vertebrates to perform this task. Birds and bats are the most important vertebrate pollinators in much of the world today. Primates often feed on flowers but typically they wolf down the whole flower, nectar and all. Destruction, not pollination, is their signature. In Madagascar, however, some lemurs still play what may be an ancient primate role as pollinators because nectar-feeding birds and bats are in short supply. Members of at least one species have shown themselves to be pollinators *par excellence*. Ruffed lemurs studied on Nosy Mangabe spent almost a quarter of their feeding time on nectar, and only fruit was a more important dietary item over the course of a year. Animals moved from flower to flower, licking as they went, while the flower itself remained intact. Pollen grains stuck to their whiskers while their faces were buried in a flower, and were transferred from flower to flower and tree to tree as they continued on their daily round.

The living lemurs also have an unusually prominent role as seed dispersers. Seed dispersers play a vital part in the life cycle of many tree species, and the fruit's flesh is their reward. In the canopy of other rainforests in the world, birds, bats and primates are the most common fruit-eaters, or frugivores. Some are seed predators, cracking open and destroying seeds buried inside the fruit. Others swallow seeds whole, transport them a considerable distance during passage through the gut and then deposit them ready to germinate, with a package of home-made fertiliser to help. Germinating at a distance from the parent tree reduces both competition and infections associated with high density.

With only seven species of frugivorous birds and three of frugivorous bats, Madagascar's trees are notably short of winged frugivores to disperse their seeds. Several lemur species have come to the rescue, playing an even larger role in seed dispersal than primates in other regions of the world. This has been closely studied in Ranomafana National Park. Five of the 12 lemur species there eat large quantities of fruit and transport seeds in their gut well away from the parent trees – beyond 15 metres, considered the 'high risk' zone, and often a distance of over 100 metres. The fruit of some tree species is far too big to fit between the jaws and be munched up by any living lemur, however. Large, now-extinct lemurs probably ate these fruits and dispersed their seeds, and the likelihood that the 'orphaned' tree species will survive far into the future is sadly in doubt.

*

Driving south from the central highlands the descent is steep and dramatic, but the sense of returning to sea level is temporary. The road soon climbs again as it heads toward the southwest and the high plain of the Horombe, stretching to the horizon. Mining and farming encroach today, but for many years when I crossed the plain it was a great, solitary place. More than anything else, the sky was splendidly on display – shimmering bleached-out blue or festooned with racing puffs of cloud in the dry season, glowering with dark storm clouds or spitting out lightning in the wet season. Palm trees studded the plain's western expanses, but an unbroken sea of grass scattered with termite mounds flanked the flat, straight, unpaved red road for much of the way. The mounds were the only vestiges of animal life visible, except for occasional herds of cattle. They would appear in the distance, grazing or making their way to some unknown destination, always accompanied by one or two men with spears trudging stoically alongside them. But that would be it. There was a lot of time to think. Long before the broad environmental history of Madagascar began to preoccupy me, I was always struck by the emptiness of the plain and would wonder what animals there might have been to see in the past.

The absence of grassland specialists in Madagascar would not be surprising. Around the world, grassland evolution and the evolution of grassland faunas did not always happen at the same time. In South America, animals specialised to feed on grasses evolved well *before* there were extensive grasslands. In North America and western Eurasia, grazing specialists evolved several million years *after* grasslands expanded. The

configuration of climate, soil type and fire was evidently sufficient to drive the expansion without the help of animals.
Perhaps there were no specialised grazers and browsers in
Madagascar either. In the early twentieth century, sub-fossils
discovered at Ampasambazimba in the central highlands were
judged to be from animals adapted to live in trees, yet the
area where they were found is grassland today. It must have
been forest when they were alive, went the argument at the
time. But recent re-analysis of these finds suggests they
belonged, rather, to animals adapted to life in a mix of forest,
woodland and grassland. Perhaps Madagascar had an open
country fauna after all. Research on the tolerance of some
endemic grass species for grazing by modern-day cattle
described in the previous chapter certainly hinted at this.

The only grazing mammals from Africa to reach Madagascar
as far as we know were hippopotamuses, which made their way
there in just the last few million years. If there was to be a
grassland community, the island mostly had to improvise with
animals already there. We could expect the results to be odd,
and they are. William Bond and his colleagues included giant
lemurs, giant tortoises and elephant birds on their list of prom-
ising candidates for an ancient grassland community 'above
ground', but did any of them feed heavily on grasses in reality?
The sub-fossil remains of these extinct species are the main
source of evidence, and their living relatives in other lands
provide further ideas about what they likely ate.

An animal's teeth, jaws and skull offer clues to what it eats.
Tooth shape and enamel thickness, microscopic scratches left
on the enamel by plants, and areas of bone where chewing
muscles are attached all reflect an animal's dietary habits, and
in living species these physical clues can be confirmed by direct
observation of feeding behaviour. For extinct species, matching

up features of their teeth and bones with those of living species with known diets is the best that can be done. This approach works for giant lemurs but, of course, far less well for animals without teeth – like giant tortoises and elephant birds. In recent years, however, carbon isotope analysis has been added to the toolkit for dietary reconstruction.

We have already encountered carbon isotope analysis as a way of tracing the abundance of C3 and C4 plants in the past. The distinctive isotope ratios in the remains of different kinds of plants are, in turn, reflected in the remains of animals that fed on them, and ratios detected in sub-fossils offer clues to what these animals ate when they were alive. The most commonly analysed material is collagen. This structural protein is present in many body parts, and has been extracted from giant tortoise carapaces, elephant bird eggshell and bone, and the bones of giant lemurs and hippopotamuses. Differences between species in digestive physiology and the rates at which they absorb traces of C3 and C4 into their bones muddy the isotopic waters. Animals are not quite what they eat. Some species absorb traces of C3 plants at a higher rate than C4, so that analyses consistently overestimate the proportion of C3 food in their diet compared to what they actually ate. The quandaries of interpretation do not end there. If isotopic clues point to the presence of 'grazer-browsers', did they live on tree- and shrub-scattered grassland or move seasonally between open grassland and forest? Living species on other continents provide examples of both, and other variants as well. The cautions notwithstanding, carbon isotope analysis is an important source of clues.

For this, first you need bones. I had been out in the forest at Bezà all morning, the sun was high in the sky, and it was hot. Walking back into the camp, I heard a spade hit the hard ground with a dull clunk. Laurie Godfrey was trying to dig a grave, and the ground was resisting. Natural history museums commonly have 'bug rooms', where beetles feed on the cadavers presented to them and yield up bones cleaned bare a few weeks later. The museum at Bezà Mahafaly does not have a bug room. It has Laurie. Whenever Laurie visits and recently dead or decomposing animals are found in the forest and brought back to camp, she strips the flesh off the bones and prepares them for study and analysis. It is a gruesome, time-consuming task, and only Laurie is brave and patient enough to do it. She is one of the most determined people I have ever met, and true grit is among the qualities that have made her arguably Madagascar's greatest expert in 'reading' teeth and bones for clues to the diets, locomotion and life history patterns of extinct lemurs. On this particular occasion, time was running out and she was behind the museum at work on Plan B – burying a *sifaka* cadaver to let nature do its flesh-stripping work, after which the clean bones would be dug up for future study.

Most of the 17 or so extinct species of giant lemurs lived in a variety of climates and forested habitats the length of the island. Wherever they lived, their teeth indicate that for the most part they ate leaves, fruit and seeds. Sloth lemurs ate all these plant parts, koala lemurs specialised on leaves, giant ruffed lemurs on fruit, and giant aye-ayes seem to have eaten a mix of fruit and insects like their smaller living relatives. A diet of foods from trees and shrubs certainly suggests forested habitat, though it does not rule out woodlands.

The *coup de grace* for most giant lemurs as candidate occupants of open habitats comes from their skeletal adaptations. They were adept climbers. Sloth lemurs were specialised hangers, suspending themselves under branches with short, stout limbs and long curved fingers and toes in order to reach their food. They include the largest of all lemurs, 200 kg *Archaeoindris*, which must have made its way slowly and ponderously indeed through the treetops. The biggest of the koala lemurs, the size of a female gorilla, were no lightweights either, and they too were slow climbers with great pincer-like feet. Giant ruffed lemurs and giant aye-ayes walked on all fours, large branches their highways, like their smaller living relatives. All these animals surely came to the ground from time to time, if only to go from tree to tree, because they were certainly too big to leap. But moving on the ground would have been slow and awkward for many species, particularly those with very long fingers and toes to trip over.

Monkey lemurs were different. Baboon-sized and baboon-like in anatomy, they had the short fingers and toes typical of animals that regularly travel on the ground, and their jaws and teeth also stand out from other extinct lemurs. There are two genera in the monkey lemur family. One is *Archaeolemur*, with two species. Their jaws were strong, their teeth thickly enamelled and, in adults, heavily pitted and scratched. Hard objects in their diet made these pits and scratches, perhaps as they cracked open seeds and nuts like capuchin monkeys in Central and South America today. Recent isotope analysis of a sub-fossil bone from Tsaramody, a high elevation site in the central highlands, adds a further, intriguing clue, for the owner of this bone evidently fed on a mix of C3 and C4 plants. The other genus in the family has only one species, *Hadropithecus stenognathus*. Alone among the extinct lemurs, it may have lived in open country in at least some areas. It has been likened to gelada baboons living on grasslands in Ethiopia, and has variously been called a grazer, seedeater, small-object feeder and hard-object feeder. None of these labels actually fits all the evidence, and recent isotope analysis of sub-fossils from the south and central highlands suggests a further dietary link – to succulent plants, which dominate southern landscapes today. For now, a question mark still hangs over *Hadropithecus*: how baboon-like really were they?

One more candidate grazer, the Malagasy relative of the African common hippo, presents itself among the extinct mammals. During my excursion with grassland experts in 2014, we clambered out of the car to stretch our legs when the road crossed a spring-fed pool. The grasses alongside the pool looked uncannily like the 'lawns' created by common hippos in Africa, remarked the experts. These hippos spend their days in the water or mud of lakes or rivers, but they actually eat little aquatic vegetation. Emerging on land at dusk, they spend the night grazing, up to several kilometres from their daytime retreat. This produces lawns with very short grass that 'would do credit to a golf course'. Untouched grasses nearby grow tall and lose their tastiness, and grazing grounds are mosaic-like as a result – lawns interspersed with unmown areas.

Cattle, not hippos, created the lawn beside that spring-fed pool in the southwest, although common hippos presumably brought their grass-eating habits with them across the Mozambique Channel. Isotope analysis of a sub-fossil hippo bone from Tsaramody in the central highlands signals that its owner ate a good mix of grasses and other plants but, curiously, grass was not a dominant item in the diets of other hippos examined so far. Madagascar's hippos surely played a role in the island's grazing and browsing community – but they rolled with the punches, it seems.

The remains of giant tortoises have been found in the south, west, and central highlands. No one argues that they climbed around in trees, but did they live on the floor of forests, woodlands or open country, or some combination of all three? And what did they eat? Mostly they chomped away on woody plants, according to isotope analysis, although the remains from two sites signal a mix of grass and woody plants or even a pure grass diet. The results do not make a clear case for one type of habitat or lifeway over another. A plausible scenario comes from a different source of evidence: the living, close relatives of Madagascar's giant tortoises on the island of Aldabra.

Aldabra is home to the only remaining wild population of giant tortoises in the Indian Ocean. On this small coral atoll 400 km northwest of Madagascar, the estimated biomass (combined weight) of tortoises may be greater than that of all the herbivores taken together on Africa's grasslands. Large body size and high population density are responsible for this feat. Individual tortoises are big (weighing over 20 kg apiece and sometimes over 200 kg), and a 1979 estimate put their number at around 150,000 individuals, more than half of them in an area of just 33 km².

Aldabra tortoises grazing (photograph by Willem Kolvoort/naturepl.com)

Aldabra is a land of scrub and grasses scattered with trees. Tortoises feed on grasses, sedges and small herbs, creating 'tortoise turf' – the tortoise version of a hippo grazing lawn – and turn seasonally to shrubs. Their lugubrious lifestyle helps make high densities possible in an unpromising landscape. They eat far less than expected for animals their size and, when food is scarce, they eat less still and stop growing or laying eggs. Giant tortoises do not stray far from trees because they have an over-heating problem, their temperature soaring if they are out in the sun too long. In the 1980s, researchers found a clear grazing cut-off about 300 metres from the nearest shade that marked the extent of tortoise forays, leaving large areas of treeless grassland with abundant potential food unvisited. Conversely, in areas near good tree cover they fed for hours.

This has consequences. When 20 or 30 tortoises regularly shelter in the shade of trees, their trampling clears the under-growth, ploughs up the soil and exposes the trees' roots. Gradual soil erosion by wind and rain follows, exposed roots are debarked, and trees die. Browsing and trampling prevent new shrubs and trees from growing and, with time, thickets and grassy woodland are converted to open grassland. With deaths from heat-stress increasing as shade diminishes, a crash in tortoise numbers ensues. Freed from assault, trees and shrubs can grow again, triggering another period of tortoise population growth as they once again find shade. This cycle has yet to be observed in full, but the evidence assembled by the tortoise-watchers of Aldabra makes an impressive case for it. Aldabra's giant tortoises provide a tantalising scenario for Madagascar, with a biomass exceeding that of herbivores in Africa, a diet of grasses and shrubs, and a role in landscape transformations between grassy woodland and treeless grassland in the space of a few decades.

Madagascar's extinct elephant birds do not have a population of well-studied close relatives living on a nearby island to consider as a model, unfortunately. As many as four elephant bird species roamed the central highlands, south, and length of the island's western flank. Although they constituted a substantial biomass of herbivores, probably second only to tortoises, research on their ecological roles began only recently. Clues to their diet come from isotope analyses of bone and eggshell, the habits of African ostriches and, sometimes, characteristics of the vegetation itself.

I was plodding through spiny forest behind William Bond. He stopped, but not to squat down and examine grasses this time. Instead, he leaned over a shrub along the path, grabbed the tip of one of its branches between his teeth, and swayed back and forth. The bendy branch swayed with him. 'See,' he exclaimed between clenched teeth, 'it's a springy forest not a spiny forest!' William was making a point about plant defences. In the past, elephant birds browsed in these forests. Lacking teeth, they could not bite through wood. Instead, they used their beaks to pull on succulent tips and break them off that way. William was demonstrating that this shrub, like many, bent rather than broke when pulled. Its architecture, he argued, evolved to resist the assaults of browsing birds. Their slender, wiry, springy branches with a peculiar zigzag architecture make them difficult to clamp or to snap when tugged: branches simply spring back into place when released. Elephant bird meals came only from soft buds and the very tips of twigs.

Isotope analyses of elephant bird sub-fossils from the south and west signal a diet of C3 plants, supporting the idea that in those regions the birds fed mainly on trees and shrubs. But they were a diverse group of species and *Aepyornis hildebrandti*, intermediate in size between its bigger and smaller Malagasy relatives, was doing things differently as it roamed the central highlands and, perhaps, further afield. The birds of this species were grazers, according to isotope analysis, with around half their diet coming from C4 tropical grasses. It is an exciting discovery – yet not altogether surprising, based on the habits of ostriches living in a variety of grasslands in southern Africa.

Even though not the closest living relative of *A. hildebrandti*, ostriches are of a similar size, and body size is an important determinant of diet. Small-bodied animals have high energy requirements per unit body weight and low total requirements. They tend to eat easily digested foods, which do not have to be abundant since they are not needed in large quantities. Large-bodied animals have lower energy requirements per unit body weight and can afford to process food more slowly, though their total food requirements are high. Their food must be abundant, but not necessarily easy to digest. Ostriches fall in the middle of this size range, and their reputation for eating more or less anything is ill deserved.

Ostrich beaks are actually quite soft and they prefer soft plant parts, seeking out new growth – green grasses, small herbs and succulents. If none are available, they turn to shrubs and the leaves and shoots of woodland trees within reach from the ground. Reminiscent of the 'springy forest' in southern Madagascar, the shrubs have physical defences and ostriches eat only green, non-woody parts. Ostriches tolerate a wide range of temperatures and do not have the stay-close-to-shade constraint of Aldabra tortoises. They are picky eaters, however, and with plenty of competition for food from grazing mammals they reach nowhere near the density of Aldabra tortoises. During the breeding season, ostriches congregate in herds of up to 50 individuals, but they forage alone or in pairs for much of the year.

Monkey lemurs, hippos, tortoises and elephant birds were all big, striking creatures, and it is easy to get caught up in the drama of their lives. As a practical matter, large animals are usually easier to find – living or dead – than small ones. Perhaps as a result, research on the ecology of Madagascar's grasslands has yet to pay much attention to small creatures that spend most of their lives out of sight underground or scurrying around in the ground litter. This is a grave oversight.

The biomass of termites in African grasslands may be about the same as that of large mammal grazers and browsers. This rather astonishing possibility alone suggests that termites, industrious decomposers and promoters of variety, are as important to grassland function as more visible members of these communities. Termites spend their days fetching vegetation, which concentrates nitrogen, phosphorus and organic matter in their mounds. They churn the soil, mixing clay with soil too sandy for tunnels or softening up clay-laden soils with sand to make them easier to excavate. Excavations also help mounds retain water better. All this makes termite mounds like little gardens, nurturing a diversity of plants with nutrients and water amid a sea of grasses.

Termites evolved more than 100 million years ago, though there is no evidence that they were present in Madagascar before the break-up of Gondwana. Today the island has three distinct lineages, probably representing three separate arrivals, and five of the 23 genera identified so far within them have species associated with grasslands and woodlands. At least one, *Microtermes*, is a genus of grassland fungal farmers that cultivate fungi inside their nests. They fetch food for their fungi and take good care of them. The return for this labour is that fungi are the farmers' main source of food. Each needs the other to survive. This 'transition to agriculture' seems to have happened only once in termite evolution, in the rainforests of west-central Africa. From there, fungal farmers repeatedly colonised grasslands in East

Africa, where they make a big contribution to decomposition processes and may even be the predominant decomposers. The fungal farming ancestors of *Microtermes* crossed the Mozambique Channel about 13 million years ago, bringing their fungi with them. The role of their descendants and that of other termites in shaping and maintaining Madagascar's modern grasslands has not been studied, but the long presence and adaptation to grassland habitats of *Microtermes* suggests that it was important.

Termites have been a major presence in my life in Madagascar. Early memories of doing battle to prevent termites demolishing the wooden posts of my hut in 1970 are overlain by perennial struggles to prevent them demolishing the wooden beams of our grander house at Bezà Mahafaly. I have eaten them with relish, fried. On walks through grasslands, I have smashed termite mounds to harvest hard chunks of earth alive with wriggling larvae for the delectation of village chickens. On one occasion I admired piles of ochre-red, indented lumps of earth outside a restaurant, while thinking of beautifully incised bowls I had seen the month before in a potter's studio in New Mexico. But no, I learned, they were not an outdoor sculpture, they were the remains of termite mounds on their way to being mixed into mortar or turned into an absorbent layer for an outdoor long-drop WC under construction in the yard. Now that I understand termites are another strand of evidence in the search for Madagascar's ancient grassland community, I feel remiss for repelling and eating them, and feeding their larvae to chickens over the years.

One other group of insects deserves a brief mention in grassland discussions: ants. Ants make their nests and forage in the soil and the rotting vegetation that litters the ground in forests. Like termites, they are important decomposers. A survey of ants in forest and adjacent grassland in the central highlands turned up almost entirely different species in each, with most of the grassland species likely endemic to Madagascar and neighbouring islands. Under an alternative scenario, ants made their way out into grasslands recently and, due to the vagaries of collecting, have simply not been found in their original forest habitat yet. This seems unlikely and, like termites, some ant species are probably important contributors to grassland function. They are also reminders of how much there is still to learn.

All in all, Madagascar may actually have done quite well with a limited stock to draw on as a 'starter kit' for an open country community. The primary credit in my view goes to giant tortoises, elephant birds and monkey lemurs above ground, with hippos piling in during the last few million years, and to termites, ants and other insects below ground. There are anomalies and puzzles, to be sure. For example, although Madagascar has endemic bird species that do well in open country, like larks and sand grouse, the diversity of grassland bird species 'ought' to be higher. Still, the balance of evidence is that grazers and browsers helped maintain open country in Madagascar far into the past.

What did these landscapes look like? Miombo woodlands have been proposed as a model for the central highlands. Grasses dominate the ground layer and trees up to 10 metres tall provide plenty of shade in these woodlands, which cover great swathes of southern Africa today. They are certainly a good starting point for imagining open country in highland Madagascar, with ponds, marshes and lusher vegetation nestled on valley bottoms. In the distance, groups of elephant birds and monkey lemurs move along, grazing and browsing in leisurely fashion, a veritable carpet of giant tortoises seems barely to move at all, and hippos snooze in ponds, awaiting their nightly excursion to feed on grasses growing on higher slopes. Meanwhile termites and ants go busily about their business largely out of sight.

Miombo woodland in southern Africa (photograph by Marco Zanfer)

The model is *only* a starting point, however, for it does not take account of the varying relief, climate and soils in Madagascar that likely supported different kinds of open country. A single scenario also misses the complexity and dynamism of grasslands through time. Even in the absence of climate fluctuations, animals and fire regimes drive change, species above ground share the habitat with species below ground, everything is related to everything else in tightly integrated systems, and a single change in one component can trigger a rapid cascade of consequences. Complexity and dynamism figure rarely in discussions of grassland in Madagascar, seeming beyond reach when the evidence is still quite sparse. Yet they are 'good to think with'. Instead of assuming stability and explaining change we might do better to assume change and explain stability.

*

141

Exploring the evolutionary origins of 'oddness' has taken us from pollination and seed dispersal to differing types of vegetation, from weather patterns to the sweepstakes route across the Mozambique Channel, from animals that spend most of their existence as eggs to an idiosyncratic community of open-country critters. As research gradually brings the island's environmental history into focus, the oddness recedes. This does not lessen the wonder of what nature has wrought in Madagascar over the last 66 million years, but it does begin to 'make sense'.

It does not explain the extinction of all the largest animals on the island over the last few thousand years, however. Does some natural upheaval account for that? I described the Horombe earlier in this chapter, but now confess that my picture of the high southwestern plain was incomplete. I was cheating, you might say, describing it only as it was on certain days. On other days, late in the dry season, there was a strong smell of smoke in the air, a brown, smoky haze hovered in the distance, spirals of grey smoke rose from the plain, and areas of black, burned grass created a patchwork of colours across it. People set these fires, burning old grass to encourage new growth and provide fodder for their livestock. Are they, and not natural upheavals, responsible for the recent extinctions? It is time to bring people into the picture.

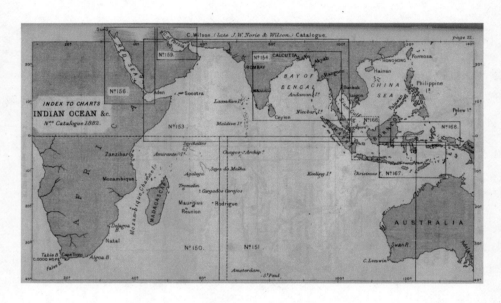

Indian Ocean and surrounding lands
(C. Wilson 1882)

CHAPTER 7

Human Footprints

It is difficult to hold millions of years and a single day simultaneously in mind, and a further 'time conundrum' arises as we approach the present. Ancient time is measured in units of millions or, at best, thousands of years because that is the closest available dating methods can take us. The recent, brief time span of human history is different, with far more precise dating methods and written records for the last several hundred years. Centuries are an everyday way of measuring and thinking about human history, and I use them here (eighth century, ninth century and so on) to locate events in time after the start of the Common Era (CE) two thousand years ago.

Chantal Radimilahy and I were sitting in the shady courtyard of a restaurant in Antananarivo on a warm summer day in January 2017. Chantal is one of Madagascar's few archaeologists, the lead excavator of Madagascar's earliest city, and retired director of the Museum of Art and Archaeology. Ours is a city friendship, and I always find it hard to picture my elegant, silver-haired friend hot and dusty in boots and jeans, wielding an excavation trowel in the middle of nowhere. That changes when she starts telling stories of life in the field. They often involve Bob. On this occasion, it was about the time when Bob suddenly vanished from the side of a lonely dirt road at night in the northeast. 'We were sure he was dead, but then he stood up'. Chantal beamed. He had stepped off a bridge, it turned out, fallen several metres onto rocks below, and landed in one piece.

Turning from nocturnal dramas, I asked Chantal about the history of archaeology in Madagascar and why so often we still seem left to guess about events in the past. She contemplated her salad as she pondered my question. The Malagasy historian Raombana made the first recorded archaeological observations in Madagascar in about 1835, she began, describing the ruins of a town in the central highlands. After that, little happened until the French archaeologist Pierre Vérin began work in the 1950s, and in 1970 established the Museum as a centre for research. He did so in the teeth of opposition from colleagues in France. There was no need for archaeology, he was told – written records already documented the history of the Malagasy people quite adequately. Vérin showed how wrong that assertion was, and Chantal, her colleagues and the students they have trained continue his efforts. Yet the ranks of Malagasy archaeologists remain thin in the face of the island's vast landscapes and much remains to be learned.

A few bones with cut and chop marks apparently made by stone tools 10,000 years ago are the first sign of people, but evidence of human presence remains thin for many thousand years thereafter. That changes abruptly with traces of hamlets and small villages in the eighth to tenth centuries along the coast, quickly followed by settlements in the interior, trading ports, towns and cities from which multiple chiefdoms and kingdoms were forged and, in the eighteenth century, a single kingdom that held sway over the central highlands and asserted itself more widely as well. Although people likely arrived in considerable numbers during the period when coastal villages sprang up, there was no single defining moment of settlement. People found their way to Madagascar at different times, chasing opportunity, fleeing trouble or simply looking for adventure. They came ashore in many places, encountered very different environments, and made a living in the new land in a variety of ways. Each region has its own history, and research in some areas, notably the east, has barely begun.

Why did a big island close to the African coast, with varied landscapes and abundant food, not attract people earlier? Humans crossed wide stretches of water to reach Australia at least 45,000 years ago, yet the only major landmass in the world on which people set foot more recently than Madagascar is New Zealand. It may be because Madagascar was 'hidden in plain sight', with winds and ocean currents making it a hard place to reach for people in canoes or small sailing boats. So where did people come from? Debris excavated by archaeologists reveals a lot about their lives once they arrived but very little about their homelands. Clues come from other fields instead – from oceanography, linguistics, genetics and botany, to naval, political and economic history and accounts written by early European visitors or told by Malagasy people themselves. As we shall see, the cultural unity of Madagascar was created from great diversity.

There are no clues to tell us what it was like to come ashore and turn an unknown place into home. Imagine canoes with foragers aboard, scanning the landscape as it loomed up ahead.

Imagine boats laden with trade goods, their owners eager to sell or exchange wares in return for new, exotic items. Imagine boatloads of men, women and children in search of a new life, with an assortment of tools, plants and animals. What a bustle of activity, excitement and apprehension must have accompanied these arrivals, what setbacks and successes must have followed. The evidence cannot capture all this, but there is a lot that it can.

*

Foragers, people who depend on wild plants and animals for food, were the first to arrive. Signs of their presence come from cut-marks on the surface of two sub-fossil leg bones belonging to an elephant bird discovered beside the oddly named Christmas River in the southwest. Stone tools used to slice meat off the bone likely made the cut-marks. The bones themselves are dated to about 10,000 years ago, using radiocarbon or C14 dating.

Carbon has already made several appearances in this book, and returns now in the most widely used technique for establishing the timeline of human settlement. C14 is a naturally occurring, unstable radioactive isotope of carbon, and the terms C14 dating and radiocarbon dating are used interchangeably. It is helpful to understand the principles of this dating method – and the *caveats*. Plants naturally incorporate stable carbon and unstable radioactive carbon from the atmosphere, maintain them in their tissues in proportions similar to those in the atmosphere, and pass them on to animals through the food chain. The processes maintaining these proportions stop when a plant or animal dies, and unstable radioactive carbon starts decaying to stable carbon at a constant and measurable rate. The ratio of unstable to stable carbon in the remains of a plant

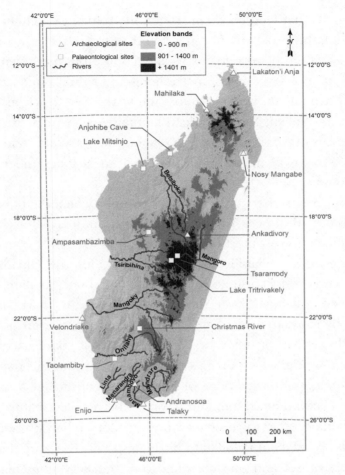

Archaeological and palaeontological sites mentioned in the text
(map by Herivololona Mbola Rakotondratsimba)

or animal tells how much time has elapsed since it died. Remains can be dated back 40,000–50,000 years using this 'clock'. In remains older than that, there is too little radioactive carbon to measure.

C14 dating sounds simple, but in practice it is not. Living plants sometimes incorporate ancient carbon, and pass this on to animals that feed on them. When the remains of either are analysed, this produces an estimated date older than it actually is. In addition, plant remains usually come from a segment of sediment core several centimetres long and so the radiocarbon date is actually an average across the years over which its raw material – pollen, leaves or dead animals – accumulated. The admixture of older material, averaging across years, and the 'plus-or-minus' of the dating method itself make for considerable uncertainty.

Today, researchers use multiple estimates whenever possible and disregard those considered unreliable, but it is well to keep in mind that even the best C14 dates may be 'off' by several centuries. The reliability of age estimates for the Christmas River chop- and cut-marked elephant bird bones is high, yet questions still hover over these bones. Were the marks *really* made by stone-tool-wielding foragers? There is no sign of the tools themselves or any other vestige of people. Bones acquire marks in several ways. A rock's sharp edge can produce a natural mark on a long-buried bone or the inadvertent slice of a spade during excavation an 'unnatural' one. The detailed features of an incision visible to the naked eye or, more reliably, under a microscope make it possible to distinguish between a mark due to these causes and one resulting from butchery immediately after the animal's death. Only features visible to the eye were used to diagnose marks on the Christmas River bones. The record falls frustratingly silent for several thousand years after this first 'sighting', apart from an apparently cut-marked leg bone from another elephant bird reliably dated to about 6,000 years ago. Cut-marks on an assortment of lemur and hippo bones suggest butchery too, but their age is either uncertain or falls predominantly within the last two thousand years, and none have been found with other signs of human presence. With uncertainties about the ages of bones and causes of cut-marks, and no associated archaeological evidence, more discoveries are needed to establish clearly when foragers first began to visit or live in Madagascar. The earliest glimpse of a place where people camped is a rock shelter, high up a steep slope in a canyon in the Montagne des Français. It is a modest place, a stretch of flat, sandy ground under a great rock overhang. After looking at many other rock shelters in the canyon, in 1986 Bob and his Malagasy colleagues decided to concentrate on this one. They named it Lakaton'i Anja, Anja's rock shelter, after the younger daughter of one of the team members, Elie Rajaonarison – who set me straight about the missionary zeal of conservationists (see Chapter 1).

The team excavated the site over several field seasons, carefully examining successive layers of sandy soil. They sifted every bucketful, cleaned the residues with acids, and sorted them wearing jeweller's spectacles. The work was slow and painstaking, but it produced treasure – stone tools, mixed in with the grit and rubble in the sieve. They were the first stone tools discovered in Madagascar, and they were old – 4,300 years old, according to the sand grains around them. Sand grains? There were no sources of carbon for C14 dating in the oldest layers excavated, and so the team instead used a high-tech method recently added to the dating toolkit, called optically-stimulated luminescence (OSL). OSL analysis determines when excavated grains of sand were last exposed to sunlight, thereby providing an estimate of the age of objects embedded amid them. Termites are a hazard for all dating techniques, because they churn the soil while tunnelling and carry remains up and down. Careful records of traces of termite activity during excavation gauge the likelihood of this, and the balance of evidence supports Lakaton'i Anja's early date.

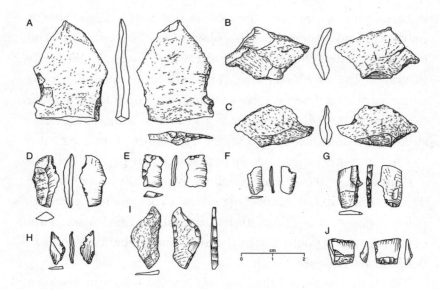

Stone tools discovered at Lakaton'i Anja (Dewar et al. 2013)

Some of Anja's tools were tiny, razor-sharp flakes that would have made lethal hunting weapons when mounted together along a wooden shaft. Making these flakes took skill, first in preparing a large stone to produce small blades when hit in just the right place with just the right force, and then in pressing tiny flakes off the blades. Other tools were of the 'smash and select' variety, produced by shattering lumps of rock too small for special preparation and selecting flakes from the debris.

Along with these tools were the leftovers from long-ago meals – bones of animals hunted in the canyon and bits of snail shell. People must have lugged food up from the coast a few kilometres away as well, for there were fish bones, sea anemone spines and remains of shellfish. The first people camping at Anja were foragers, and they travelled light. They also ranged far. Most of their tools were made of flint from a nearby source, but a few pieces of crystal quartz originated at least 120 km away. People returned to Anja at intervals for several thousand years. They continued to use stone tools, but the world was changing. Fragments of locally made pottery join the mix in younger layers and, by the twelfth century, elaborate porcelain from the Arabian Gulf and China. The people camping at Anja by then almost certainly had a base elsewhere, a hamlet or village tucked in one of the many valleys in the massif or on the coast nearby. The old days, when foraging was the only way of life, were gone and other activities, like trading, were growing in importance.

Bob, the children and I visited Lakaton'i Anja together in 1991. The day was hot but the breeze was cool as we scrambled up a path barely visible amid the dense vegetation. We climbed and climbed, and I wondered how on earth Bob and his colleagues had found this place. It seemed no different from all the other rocky slopes pocked with overhangs that we passed on our trek up, and this one could not be seen from the valley floor. I looked at the shelter, cool, dark and inviting, and thought about the shelter I had slept in years earlier on the southern tip of the island, in the forests of Andohahela National Park. Did the people who slept here feel as cosy, dry and safe as I had done then? A stream running along the valley floor meant that there was water nearby, the dry forest offered plenty of firewood, and breezes off the sea made the heat tolerable. It must have been a pleasant place to camp, and I could imagine why people would return over the centuries. I never saw what Bob did during the half of the summer when I was home taking care of the children before we swapped places and I set off for Madagascar myself. Our hike that day gave me a better understanding of why he loved doing it.

The trouble with Anja is that nothing like it has been found elsewhere in Madagascar. My prediction is that more early sites with stone tools will be discovered now that it is clear they are worth looking for. The closest to Anja in time is a cluster of sites at Velondriake, an arid stretch along the southwest coast. Small forager groups began camping in rock shelters there about 2,800 years ago, leaving behind elephant bird eggshell and seashell fragments, and traces of charcoal from their cooking hearths. There are no stone tools at Velondriake, though they are hard to find and may yet be discovered. Velondriake was occupied for even longer than Anja and, like Anja, the way people lived slowly changed. Between the fourth and sixth centuries an open-air fishing camp developed on a beach near the rock shelters and grew into a village over time. Locally made pottery appears, looking much like that from other regions of the island, and a few beads made from elephant bird eggshell and seashell show up too.

Where did these early foragers come from? Madagascar was surprisingly inaccessible. Monsoon winds favouring passage from lands to the northeast only reached the island's northern tip, and only in one season. Along the east coast, the winds and surface current flowed from faraway Indonesia. The southeast trade wind blew in year-round across vast, landless waters of the southern Indian Ocean, and to the south swirled the circumpolar current. The first humans to set foot on the island almost certainly came from East Africa, even though sailing or paddling across the treacherous waters of the Mozambique Channel would itself have been a dangerous undertaking. People living along the coast were making short trips by boat to nearby islands several thousand years ago, however, and the stone tools at Anja look much like contemporary tools along the African coast. The makers of Anja's tools were probably Bantu, a label loosely given to 300–600 ethnic groups in Africa who speak related Bantu languages. A few linguistic clues hint that a 'pre-Bantu' language was spoken in earliest times in Madagascar, and the existence of a distinctively different people in the distant past is a recurrent theme of modern Malagasy folklore and oral histories. But the evidence is thin.

People may first have arrived by accident, in boats blown by the wind or carried to shore by a stray current, or perhaps they ventured to the big island on purpose, exploring, or fleeing conflict and searching for land to forage in peace. For now, writing about these early foragers can feel more like telling a ghost story than reporting history. Happily, the people who established small coastal settlements during the eighth to tenth centuries left behind charcoal from hearths, animal bones, broken shell and an abundance of pottery sherds for archaeol-ogists to discover. Pottery sherds hold little allure for me, I confess – although they should, because they are important evidence. For one thing, style and manufacturing technique contain clues to when a piece of pottery was made, and so sherds can suggest the period when a settlement site was occupied and provide an independent way of confirming or challenging dates estimated by C14, OSL and other methods. For another, they provide clues to trading. And sometimes they speak, or so it seems when Henry Wright is around.

Henry is a distinguished archaeologist and, rumour has it, model for the father of Indiana Jones in *Indiana Jones and the Last Crusade*. He and Bob worked together in Madagascar for many years, and Henry would appear on our doorstep in Connecticut from time to time, a beret on his head, a backpack slung over one shoulder and a magic carpet rolled up on the other. On this, we flew off around the world. Henry has a stun-ning ability to spot similar production processes and decorations in pottery found hundreds or thousands of kilometres apart and transform them into insights about the settlement history of Madagascar. Sherds figure prominently in what follows!

Archaeologists have so far mapped a dozen or so hamlets and villages dated between the eighth and tenth centuries, and remote sensing suggests that many more sites await discovery. Consider two of them – one in the northeast, with lush forest and plentiful rainfall, the other in the harsh, dry lands of the south, a much less obvious place to settle. The little island of Nosy Mangabe where I had my first glimpse of aye-ayes is quite a precarious place to live, with frequent cyclones barrelling in from the Indian Ocean, but it is also warm, wet and verdant, with a protected sandy beach and rushing streams on the landward side, and settlers took a chance. Most would have been farmers, growing rice along with other crops, but there were craftsmen in the community too. Excavations turned up fragments of coarse local pottery, an assortment of minerals, and iron slag left over from smelting and forging. Pieces of pottery from the Arabian Gulf were mixed in with these items originating in Madagascar, showing that the horizons of Nosy Mangabe's settlers lay far beyond their little island. They were connected to the Indian Ocean network, acquiring goods from overseas and possibly dispatching exports – soapstone, iron and fine rock crystal.

The site of Enijo, a flat area overlooking the mouth of a river near the southern tip of the island, is a far cry from Nosy Mangabe. Covered with spiny vegetation, the region is baking hot by day and chilly by night. Perhaps the best that can be said for it is that it is less likely to be hit by cyclones. Yet people settled here too, and severe storms in recent years scattered the contents of a rubbish pile in the ancient village of Enijo across a nearby beach. Rubbish piles count as archaeological gold, for they preserve the thrown-away leftovers of everyday life, including pieces of broken pottery. The pottery sherds found at Enijo turned out to be quite different from those found anywhere

Pottery sherds from Enijo (Parker Pearson 2010)

else on the island, including sites just a few kilometres away. By far the best match is with contemporaneous pottery from the East African coast. Like the villagers on Nosy Mangabe, people living in Enijo evidently had ties overseas. Did they sell or barter goods in return for pots? Perhaps elephant bird eggs and semi-precious stones found nearby, but that is speculation. If 'ghostly' springs to mind for traces of the world of early foragers, 'scrappy' might be the adjective for traces of the farmers, fishing folk, craftsmen and traders who arrived after them.

*

Trade linked Madagascar's settlers to the world beyond its shores. The Indian Ocean trade network emerged about 2,500 years ago and flourished until the fifteenth century. Its long promi- nence was thanks to the Asian monsoon system. The system powered boats over long distances across northern reaches of the Ocean, and the seasonal reversal of winds made round trips relatively easy – easier than crossing the Atlantic, where the trade winds blow in one direction year-round. Only when sailors figured out how to cope with this did the Atlantic become a dominant centre of overseas trade. The monsoon system under-pinned trading opportunities around the Indian Ocean in other ways too, for wind direction influences patterns of rainfall and rainfall drives agricultural cycles. It was, in effect, a natural mechanism for synchronising land-based production with trans-ocean trade.

Tucked away in a far corner of the Ocean, Madagascar's incorporation into the network came relatively late. There were many trading opportunities that did not involve an arduous voyage to such a distant island. But the Mozambique Channel eventually became a busy trade highway – certainly by the eighth century, and possibly earlier. A first-century handbook of the Indian Ocean for Mediterranean merchant sailors described navigation routes and prospects for trade, listing ports and coastal landmarks. They extended along the African coast to within 1,200 km of Madagascar, and perhaps some merchants made it to the island. A round trip all the way between Indonesia and Madagascar around the Ocean's rim was a very long journey indeed, however. Under sail, and including stops along the way to trade, resupply, and await favourable winds, it would have been a two-year undertaking.

A different route between Indonesia and Madagascar was possible in principle: a one-way ticket across the middle of the Indian Ocean, riding the Equatorial Surface Current and winds that stream from east to west. The voyage of the *Sarimanok* is the most famous modern attempt to show that this route was also possible in practice. Makers of the boat for the voyage constructed it according to the plan of ancient Filipino ships: known as a double outrigger canoe, it was a single, 30-metre-long, hollowed-out tree trunk, with bamboo floats on each side and sails made from palm leaves. Everything was lashed together with cords made of bark or vines and, in line with past construction techniques, not a single nail was used.

The *Sarimanok* left Bali in Indonesia on 3 June 1985 loaded with food, water and a crew of seven Europeans, their sole navigation instruments the sun, moon and stars. On 6 August, the northern coastline of Madagascar loomed into sight. But a storm blew the boat past the entrance to the great bay at the tip of the island, and a French Navy patrol boat ended up towing it into harbour in the Comoro Islands. The *Sarimanok* set forth again a few days later, and finally arrived in triumph on the tiny island of Nosy Be just off the northwest coast of Madagascar. Its crew members were brave, mad, skilful and inept. Whatever else, they demonstrated in modern times that the Indian Ocean could be crossed in a dugout canoe with floats and sails. They also showed that, absent a French patrol boat, Madagascar is easy to miss even when close.

Winds, currents and grand boating adventures suggest what was possible or likely but not what actually happened. Several intersecting lines of evidence help with that. Today, the island-wide genetic make-up of the Malagasy people reveals an ancestry that is about 60 per cent African Bantu and 40 per cent Indonesian on average. Markers of Bantu heritage are more prevalent in coastal areas and Indonesian in the interior, although the wide diversity of people present at weekly markets in the countryside today is vivid testimony to a history of much local mixing-up too.

Based on analyses of the Malagasy language, linguists generally agree that the initial coming together of Africans and Indonesians took place during the eighth to tenth centuries. Where did they first encounter one another? Some linguists have argued that would-be settlers from Indonesia hitched a ride with traders or were themselves traders. Reaching East Africa, they mixed with the people they met there and embarked for Madagascar together. Others have made the case that Indonesian settlers arrived independently and encountered Africans for the first time on the island itself. Genetic evidence supports the latter scenario, although what actually happened was almost certainly more complex than a simple 'either/or'.

Efforts to locate more precisely the homelands of people within Indonesia have produced a grand enigma. Around 90 per cent of the Malagasy language today has origins in Indonesia, and its closest relative is spoken by the Ma'anyan. They are forest-living people in central and southeastern Borneo, with no history of seafaring, and the genetic links of modern Malagasy are much stronger in eastern Borneo and other islands in Indonesia. How the language of the Ma'anyan ended up being the foundation of the Malagasy language is a mystery indeed.

Several lines of evidence suggest that some settlers did indeed follow a direct path across the middle of the ocean, as, in our own time, the *Sarimanok* tried to do. Ocean-going boats with expert navigators were not lacking in the region. As early as the third century, Southeast Asians were building boats 50 metres long with several masts and sails, their wooden planks 'stitched' together with plant fibres. These boats transported great loads of cargo and hundreds of passengers over long distances across rough seas. Settlers could have travelled on big ships like these, on more modest craft, or perhaps aboard boats crewed by sea nomads, wanderers and expert sailors who have lived from fishing and trading in the Indonesian islands since the sixteenth century and probably much earlier.

Excavation of a 30-metre-long boat from the seventh to eighth century,
discovered at Punjulharjo on the northern coast of Java (EFEP/P-Y. Manguin)

Whatever route they took, the historical impact of cultivators from Indonesia is clear in the Malagasy language today. Malagasy words for crops long grown there are largely Indonesian in origin. Words for domesticated animals are more closely linked to Africa, on the other hand. Common sense kicks in here: stowing sacks of rice, coconuts and yams on a boat for a long journey on the high seas is one thing, managing a Noah's Ark of animals quite another. Botanical samples from eighth- to tenth-century archaeological sites along the African coastline,

Madagascar and the Comoro Islands are strikingly different from one another. Asian crops are dominant at sites in Madagascar and the Comoro Islands, whereas African crops predominate at African sites and Asian crops are absent or rare. They appear in very small quantities at most, usually in major trading ports, and probably imported by traders.

Looking back over the centuries of settlement, my guess is that some traders arrived, liked what they found, and decided to base their activities on the new island, while others dropped off passengers from Africa or Indonesia. Tips from traders may also have encouraged Indonesians to take the direct route across the Indian Ocean. Setting out without some idea of your destination is hard to imagine, and word of a big island with plenty of space on the far side of the ocean would have reassured people contemplating the long voyage.

The broad history of the Indian Ocean region hints at reasons why would-be settlers left their homes in Indonesia. The period was one of environmental change, with knock-on economic effects. Between the fourth and ninth centuries the northern hemisphere became drier, temperatures fell, and the southwest monsoon weakened. In a distant echo of our own times, human activities may have caused some of these effects. Economic activity slumped around the Ocean, and disease, travelling via trade routes, made its own devastating contribution. They were the kind of times that led people to pick up and start again somewhere new. Political developments in Southeast Asia could be a reason too. From the seventh century, Malayans expanded their activities and gradually developed an empire ruled from Sumatra that came to hold sway over much of Indonesia. Chilling inscriptions carved into stone pillars erected on a small island just off the Sumatran coast pronounced a curse on those who refused to yield to the new master race conquering the land. I can imagine wanting to flee a master race. The modern drama and tragedies of 'boat people' may have played out in those days too.

Women and men in the modern population of Madagascar have distinctively different maternal and paternal genetic lineages, adding a further enigma to the process of settlement so many centuries ago. It has been inferred from these findings that early settlers from Africa included more men than women, whereas those arriving from Indonesia included a greater preponderance of women. How interesting it would be to understand the circumstances that drove this difference, picked up today only by faint genetic signals . . .

*

With the age of arrivals over, from the eleventh to the fourteenth century settlers expanded into the interior, the first major town rose and fell, and all the important historic economic activities of the Malagasy became firmly established: growing rice, herding livestock, fishing, smelting iron, and overseas trading.

In the central highlands, farmers set to work in the rolling hills and marshy valleys. Excavated soils contain high levels of charcoal and the pollen of weeds typically found in fields, indicating that people were clearing and planting the land. Rice was almost certainly among the crops they grew. They lived in little villages nestled in valleys, each village surrounded by a shallow ditch – likely a way to prevent cattle from wandering in. At one of these, Ankadivory, a few fragments have been found in the style of pottery from the southeast coast at that time, one of several indications that the villagers moved up to the highlands from the east. Local potters were developing their own distinctive style too, and most of the pottery they made was quite different from the southeast or any other region of the island. Yet fragments of wares imported from the Near East and Asia tell us that these inland settlers were connected to the Indian Ocean trade network.

In the south, archaeologists have mapped 58 settlement sites in the four great river systems of the region, the Mandrare, Manambovo, Menarandra and Linta. Between them, the sites supported a dense concentration of people. Andranosoa sits on the bank of a river about 80 km inland. It was a big place for its time. A discontinuous embankment of loose stone ran around it, like many settlements in the region. Embankments had a defensive purpose in later times but Andranosoa's was puny, a barrier for livestock perhaps. Debris from the town was scattered over a wide area but the jackpot was an old rubbish pit, as dear to an archaeologist's heart as a rubbish pile. Pits were useful to the people who originally dug them, providing sand and soil for house construction, and places to store crops or cook food before meeting their eventual fate – accumulating and concentrating in one place the detritus of everyday life.

The contents of the Andranosoa pit are intriguing: Chinese porcelain, local pottery, lots of fish and sheep or goat bones (hard to tell apart), a few cattle bones, one small piece of cara-pace from a giant tortoise, remains of about 50 tenrecs, and a few hippo teeth. Tenrec bones are scarce at other sites in the region. Perhaps someone in Andranosoa had a particular taste for them. The giant tortoise and hippo scraps are a particular puzzle. If people regularly killed these animals for food, you would expect there to be many more leftovers in the pit. Maybe someone stumbled across the scraps on a walk and brought them back to the village as curiosities. Who knows . . . The bones of cattle are the earliest evidence of their presence in Madagascar. All Malagasy cattle today are zebu with origins in South Asia, with a distinctive fatty hump on their shoulders and high tolerance for hot weather. Although nearby Africa would be the closest source for the first zebu, genetic analysis suggests that people transported them directly from India. Cattle were introduced more than once, however, and hump-less cattle from Africa also figured among early imports.

The people of Andranosoa herded livestock, foraged and fished. Like the farmers of Ankadivory in the central highlands, they were part of an overseas trade network even though they lived well inland. Perhaps, as at Enijo, they had elephant bird eggshell, semi-precious stones or even cattle to barter or sell. Anyway, things were hopping there – unlike the fishing village of Talaky, downstream at the river mouth, where excavations yielded iron fishhooks and harpoons, lots of marine shell and fish bone, coarse local pottery, and not a single fragment of imported wares. Whether home was a cosmopolitan town or a simple fishing village, life in the south was as uncertain then as it is now. The regional population declined dramatically from the fourteenth to the seventeenth century, likely due to a succession of devastating droughts. Only in the twentieth century did as many or more people live in the south again.

Heading up the coast to the northwest brings us to Mahilaka, Madagascar's earliest major trading centre. Chantal Radimilahy, whom we met at the beginning of the chapter, led the excavation of this important site. With the most equitable climate along the whole western flank of the island and access to anchorage in a sheltered harbour, Mahilaka was bigger than any contemporary town in Madagascar or along the east African coast between the twelfth and fourteenth centuries, with 3,000–5,000 residents in its prime. Within its walled 45 hectares were mosques, a number of large, well-built homes, and workshop areas for iron smelting and forging, soapstone carving and glassmaking. An extensive network of small villages and hamlets around the prosperous town supplied food. Excavated grains of rice, coconut shell fragments, cow and fish bones and seashells speak to a varied and tasty diet, and also to the industriousness of its suppliers.

Mahilaka was the primary exit and entry point for traders in Madagascar and overseas. From its hinterlands people brought rice, cattle, timber, aromatic tree gums and basketry, and a deforested corridor visible today may be the vestige of a well-worn route taken by people transporting these products and loads of rock crystal, soapstone, gold and iron from the mineral-rich northeast. The biggest concentration of waste rock crystal found anywhere between the eleventh and fourteenth centuries is at Mahilaka. The purity and translucency of Madagascar rock crystal are ideal for glassmaking, and it was probably used for exquisite pieces crafted in the Islamic world during this period.

Some of the goods carried to Mahilaka were to meet local demand, but many of them were destined for export. In return, traders bought cloth, beads, silver, pottery and fine porcelain from China. Urban life in Madagascar began there, and the presence of an elite who lived in the biggest, best houses in town is the first clear sign of social hierarchy on the island. The town was largely abandoned around the end of the fourteenth century, and other northern ports took on its role in overseas trade. The reasons for its downfall are unclear. Perhaps the surrounding area became so depleted it could no longer supply enough food for the town's population; perhaps there was strife or sickness, or the focus of international relations shifted else-where for some other reason.

Travelling forward in time to the fifteenth century and beyond, written records supplement the archaeological evidence of human activities. Hierarchically organised communities spread

and, by the seventeenth century, the landscape was scattered
with small, dynastic kingdoms 'too numerous to count', engaged
in near-constant competition and conflict with one another.
Heavily fortified centres housing a thousand or more people
became widespread, with small settlements clustered around
them, and fancy local goods, imported goods and special tombs
marked the status of political chiefs headquartered in the big
towns.

From these many, small chiefdoms gradually emerged a few
much larger political units and one, the Merina Kingdom, came
to dominate the central highlands and assert varying levels of
control over other regions of the island. The roots of the Merina
Kingdom were in small, thirteenth-century settlements like
Ankadivory. Over succeeding centuries, these settlements multi-
plied in number, grew in size, and shifted to fortified hilltop sites
as the frequency of raids between chiefdoms mounted. The
influence of the central highlanders began to spread. Local
pottery styles in other regions gave way to distinctive styles seen
in the twelfth- to thirteenth-century pottery of Ankadivory, and
similar motifs appear on fabrics, silver and wood as well, perhaps
signalling that shared aesthetics and beliefs were increasingly
embedded in a pervasive Malagasy culture and language.

Large-scale rice production became key to Merina prosperity.
The supporting irrigation systems required major investment
together with societal coordination and control, hallmarks of a
hierarchically ordered and administered state. These developments underpinned Merina efforts to conquer neighbouring
kingdoms and control internal and overseas trade – producing
still more wealth. Trade included many craft products, from
iron tools to textiles, and cattle captured in battle that were
herded to the central highlands and then down to the east coast
for export. It also included the import of guns and the export

of slaves. Slavery has a long history in Madagascar, probably starting with the enslavement of captives by feuding groups on the island and growing into an overseas trade with Arab and then Portuguese slavers. From the mid-seventeenth century, European traders expanded the business massively and Madagascar became a major hub in the East African slaving network. This lasted until the new French colonial administration abolished slavery in 1896, and went on to establish forced labour in its place.

It is easy to imagine people's lives as quite sedentary before the advent of roads or cars, but that is a mistake. Free people moved around in Madagascar because that was how they made a living. Enslaved people moved because they had no choice. The travels of an unnamed Malagasy ex-slave recounted to a Goan Jesuit in 1614 illustrate the remarkable distances sometimes involved. Enslaved as a child in the east, his captors marched him northwest across the island and loaded him on to an Arab boat bound for the Comoro Islands. Arabia, then Mozambique and, eventually, England followed. After several years in England, he made it to Holland, somehow regaining his freedom along the way. A Dutch ship carried him to southeast Madagascar, and he walked north and went home.

A century ago, there were close to 3.3 million people on the 587,000 km^2 island. After almost 10,000 years, Madagascar was still sparsely occupied. The population is around 28 million today, but compare that to the 66 million people in France, a similarly sized area: the human population density of Madagascar is still relatively low.

*

It is hard to remember a time when Jean-Aimé Rakotoarisoa was not part of my life in Madagascar. He was Director of the

168

Museum of Art and Archaeology and Bob's primary collaborator when we met in the 1970s, and collaboration became a family friendship. Archaeologist by training, Jean-Aimé is a Renaissance man by nature, and a leading catalyst of research on the history of the Malagasy people. I have never been in the field with Jean-Aimé but we have spent many evenings together, and listening to him has been a major influence on the way I think about Madagascar. Regional histories and the links between them are widely acknowledged to be important now, but my own long focus on 'place' owes much to conversations with Jean-Aimé. Invariably, though, the talk would drift back to the conundrum of a people whose undoubted unity is paradoxically rooted in their undoubted diversity.

Overseas origins are not part of Malagasy identity today, and no discrete 'packages' of peoples and cultures have made their way to the present. To the contrary, as Jean-Aimé has written, 'The Malagasy are from Madagascar. This statement may seem a truism . . . but past studies have tried to explain and define the Malagasy largely in terms of their overseas origins . . . yet [these] play no part in the lore or identity of most rural Malagasy. The only "ancestral land" (*tanin-drazana*) they know and recognise is the island of Madagascar itself; their only ancestors are those residing in the tombs that conspicuously dot the landscape'.

Tell a Malagasy farmer that the vanilla he or she cultivates originated in Mexico, and you are likely to be looked at as if you need your head examined. That, at least, was the experience of Sarah Osterhoudt in the village of Imorona on the northeast coast near the city of Mananara in 2009. She and her husband had first come to the village as Peace Corps volunteers working to establish more direct links between vanilla farmers and US markets. By the time we met, she was completing a Yale PhD in Anthropology and Environmental Studies on the lives of these farmers and the role of vanilla within them. Along with saffron, vanilla is considered the most labour-intensive crop in the world. The moth that naturally pollinates vanilla in Mexico has never managed to establish itself elsewhere and so every flower has to be pollinated by human hands.

During Sarah's sojourn, the Imorona farmers decided to write a booklet for other farmers about the skills needed for successful cultivation. In the course of conversations about what it should say, she told them that the French first brought vanilla to Madagascar around 1840. The news provoked disbelief and lively debate. When the booklet was finally drafted, the section describing the history of vanilla left out the role of the French completely, saying only that vanilla was a 'child' of Mexico that 'brought itself' to Madagascar, and Malagasy words replaced the many terms used for vanilla cultivation that were actually French derivatives. Sarah's account is interesting not only for the farmers' resistance to acknowledging colonial subordination but also for the way they set about incorporating into a distinctive Malagasy universe a plant they had just learned was not native to Madagascar. It is a process that must have repeated itself time and again, as settlers from far-flung places became a 'people' and the island became 'their' island.

For a long time, people living in Madagascar had no name for the whole island. At least until the seventeenth century, they simply gave the name of their own and neighbouring regions when asked. A possible time to locate the emergence of the 'Malagasy people' is when the Merina Kingdom expanded its reach in the eighteenth and nineteenth centuries and declared itself the Kingdom of Madagascar. But it is unlikely the kingdom actually functioned as a unified nation state with a single cohesive economy and, besides, political unification is not everything. Close interactions among the island's occupants go back many centuries earlier, and likely nurtured the paradoxical unity rooted in diversity that characterises Madagascar today.

Remember the Arabian Gulf pottery sherds on Nosy Mangabe, East African sherds at Enijo, fine Chinese porcelain in the rubbish pit at Andranosoa, and Middle Eastern and Asian sherds at Ankadivory. Overseas trade was an early feature of settler life, and trade networks across the island would have developed to supply and profit from it. People congregating at markets to buy and sell must be able to communicate, and settlers from different homelands speaking different languages needed a *lingua franca*. The Malagasy language likely arose to fill that role. Over time, it became everyone's first language, albeit with distinct regional dialects, and it is spoken universally in Madagascar today.

Explorer and missionary reports between the sixteenth and eighteenth centuries indicate that the Malagasy language had certainly emerged by then, and possibly much earlier. Their descriptions are remarkably consistent: people spoke some form of the language everywhere except the west and northwest coast, which seems to have been a linguistic chequerboard of Malagasy dialects and Bantu languages. In the early nineteenth century, Nicolas Mayeur commented in a letter, 'I was understood everywhere'. Mayeur was a French merchant, slaver and translator, who had learned the Malagasy language as a child in the north and spent many years travelling around the island as an adult. He knew whereof he wrote.

A melting pot of people, Madagascar forged a unique material culture too. Music making is my own personal favourite instance of unity created from diversity. Some musical instruments played there today originated in Southeast Asia, others in Africa. But listen to the music played on them and the language used in song, watch the dances performed to their melodies, and what you hear and see is a wondrous cultural mixing-up. This kind of musical Tower of Babel, sitting right in the midst of the modern state of the Democratic Republic of Madagascar, has travelled far and wide. Malagasy-American jazz composer Andy Razafy wrote tunes that are now part of the American jazz canon, for example – like 'Honeysuckle Rose'. I made my own modest contribution long ago, while living in the little village of Hazofotsy, in the south. 'Where have all the flowers gone?' I sang (words by American composers sung with an 'English English' accent), to the accompaniment of a *lokanga* (three-stringed fiddle originating in Africa), while everyone else sang along in Malagasy (originating in Indonesia) and danced (originating who knows where . . .)!

*

Reinaldo Rasolondrainy was sitting at a café table in Toliara with Bob and me on a sweltering October day in 2012. His six-year-old son sat beside him, silent, still, immaculately well behaved. Nado, as he is known, had recently finished a Master's Degree at the University of Dar es Salaam on rock paintings in southwest Madagascar, and had been excavating up north with Bob earlier in the season. His ambition now was to do a PhD in archaeology at Yale University. In his other life, he is a celebrated singer in Madagascar.

Our conversation was mostly about the application process, letters of reference, evidence of language skills and so on. Nado has a smile that lights up the world and a sunny disposition to go with it, somehow rendering lighthearted even the weighty topic of university admissions. Eventually, he and Bob turned to the archaeology of Madagascar and the particular topic he might tackle for a PhD. It was very interesting, but what I remember now was the sudden ferocity of Nado's last comment:

172

'My son goes to the French school. Today he was taught that Marco Polo discovered Madagascar. That's *wrong*, and I want my work to help change that.' For a moment the smile disappeared. Almost a decade later, Nado has a Yale PhD, and the archaeology of Madagascar slowly becomes more widely known.

Hippopotamus, Hippopotamus lemerlei, *jaw bone from Taolambiby*
(drawing by Luci Betti-Nash)

CHAPTER 8

Receding Forests

It was a warm winter morning in the southwest and I was happy to be back with Joelisoa and my Bezà Mahafaly friends, catching up on the news. We were squeezed into a car together, wallowing along the pitted track to the village of Taolambiby. Joelisoa was keen for me to meet the village president, a respected elder in the community, and visit a nearby site famous for its ancient animal remains. It would have been easier to walk there, I thought.

The car came to a halt amid swirling dust outside the president's hut. Many people go by one name in Madagascar. The president's name was Eforida. We stumbled out, shook hands, sat down gratefully on a mat in the shade of a tree, accepted his invitation to drink shots of local rum, and began the usual exchange of news and gossip. The rum took the back off my throat and landed with a pleasant glow in my stomach. A small boy chased a scrawny chicken around behind me, but the chicken was faster than he was. The afternoon wore on, the conversation interesting as always. Cornered at last, the chicken was presented to us with its feet tied. It would be dinner.

Then we were on our feet again, plodding in single file behind Eforida along a path through straggly bush. Many trees and shrubs in the southern forests lose their leaves at this time of year, and all was silver, brown and grey around us. Treading in Joelisoa's dusty footprints briefly turned my thoughts to tracking. People in this region read the ground. Word of our procession of villagers, professors and scampering children would spread quickly.

For the next quarter hour Eforida led us further into the bush, the only sound a soft pad of sandals and boots. Everything seemed asleep, shut down by the midday sun. But then the path dropped steeply, and we entered a different world. It was shady and cool, and a pool of water lay below us with banks topped by green, leafy trees. A rocky gully leading from the pool was dry now, but in the brief rainy season it would fill, and, over the years, rushing water had carved a deep channel. Eforida set off along the gully, glancing at the bank on our right. He did not go far before pulling from it a big grey lump. He gave me the object, smiling. *Omby rano*, he said. Water cow. An ancient hippopotamus jaw lay in my hands. I stared at the teeth set in the massive jaw like polished white pebbles, and pondered. The landscape must have been far wetter and greener when this animal lived. When were the lush forests of the past reduced to tiny glades like the one in which we stood? Is this what caused hippos to disappear?

The question in my mind that day would have taken a different form in other parts of Madagascar. Its diverse regions have been likened to a cluster of interconnected small islands within a single big island. Each small island has a long and distinct history of its own, including different patterns of human settlement and activity over the past several thousand years. The south, central highlands and northwest are the regions where the richest evidence is available for changes in vegetation from a little before the estimated time of arrival of people up to the nineteenth century. What or who was responsible? Multiple, interacting causes are known to drive change in many parts of the world, from climate, natural fires and wild herds of grazers and browsers, to the activities of people and their livestock. Solid timelines supporting tight cause-and-effect scenarios are elusive. Even the best dating techniques narrow the time at which events took place to within a few centuries, and their temporal sequence is rarely precise enough to be really confident that *this* happened because of *that*. Last but not least, island-wide events were rare, and so evidence must be painstakingly collected and analysed region by region.

Faced with these challenges, it would be easier to take the presence of people in Madagascar and broadly contemporaneous changes in the environment as proof of human culpability and give up trying to figure out the cumulative effects of varied pressures over many centuries in many places. The sight of people clearing forest is difficult to miss today, and it is a short step simply to project this back in time. That would be a mistake. The regions visited in this chapter contain a lot of good evidence, variously embedded in sediment cores, stalagmites and archaeological excavations, and written accounts add another dimension closer to the present. Let us see what this evidence tells us.

*

The south of Madagascar still lies within the arid belt circling the southern hemisphere. It is a harsh land of frequent and sometimes prolonged droughts. Ten-thousand-year-old cut- and chop-marked *Aepyornis* bones were discovered in the southwest. Small hamlets and villages along the southern coast well over

1,000 years old were among the first on the island. The south is where the bones of cattle first appear in excavations. The current evidence suggests that people have been in the south for at least as long as any other region of Madagascar, and perhaps longer. If there is a strong link between their presence and habitat destruction, this should be the region to find it.

The deep south is too dry for lakes and so there are no lakebeds from which to extract sediment cores. Most clues to the region's climates, habitats and fire regime come from cores drilled in lakebeds in the slightly less arid southwest and a moist corner of the southeast, although a single stalagmite in the southwest has recently added a new level of detail for that area. I wrote earlier about fragments of pottery that almost seemed to speak thanks to the expertise of archaeologists, and I recently started feeling the same way about stalagmites – equally unlikely candidates for speech.

Stalagmites are formed by water dripping slowly into limestone caves over thousands of years. The water contains isotopes, and sometimes pollen grains, and these are incorporated into the stalagmites as they grow, waiting patiently for researchers with interesting questions and sophisticated technology to unlock their secrets. Oxygen isotopes provide signals of climate, carbon isotopes a record of vegetation, and uranium and thorium isotopes a way to date successive layers of stalagmite growth. Between them, they constitute a sometimes decade-by-decade record of change over thousands of years and, crucially, make it possible to track past climates and vegetation independently. If the changes in oxygen and carbon isotopes go hand in hand, climate was likely controlling what type of vegetation grew in the surrounding area. If they are out of sync, with vegetation changing under stable climate conditions, the hand of people was probably involved.

Climate fluctuations and their impact are vividly on display in several cores and the single stalagmite examined so far in the southwest. It was good hippo country 6,000 years ago, much wetter than today. But the good times did not last. Sometime between 3,000 and 4,000 years ago, conditions became increasingly arid, marshlands and ponds dried up, natural fires increased, dry forests retreated, and palm-studded grasslands spread. Crocodiles and hippopotamuses – like the one whose jaw I held in my hands at Taolambiby – declined or disappeared. Water birds disappeared too, and only generalist species survived and persist to this day.

Isotope data from the sampled stalagmite remain in sync with one another throughout this period of drying, and the synchrony persisted even after human settlement. Sediment cores likewise indicate little change in vegetation when villages were first established or in the centuries that followed. The lack of discernible human impact is not surprising. Most villages were just a few hours' walk from the coast, the population was still sparse and, with the sea nearby, it was likely easier to live by fishing or trading than farming. People would have grown crops on a small scale around their villages, but the preparation by hand of land for planting is back-breaking work. Trees must first be cut down and left to dry, a fire lit to burn off undergrowth and convert dead wood to ashes that will serve as fertiliser; finally, the soil has to be ploughed. Only then are seeds sown. Faced with such a major task, communities were unlikely to transform whole landscapes this way under climate conditions unpropitious for agriculture.

The southeast corner of the island, beyond the eastern mountain range, is considerably wetter than the rest of the south. Sediment cores from four sites a few kilometres apart track changes in climate and vegetation there over the last 6,000 years.

With precise dating techniques, the cores tell a fine-grained and quite different story from the southwest. There is no sign of the prolonged drying that gradually transformed the southwest. Short periods of drought rarely had a discernible impact and, in fact, the *lack* of synchrony between climate fluctuations and vegetation changes is striking. One possible exception is a broad decline in forest cover during a time of drought and higher sea levels a little over a thousand years ago. But people were spreading across Madagascar by then. Did human activities contribute to the environmental shift? Probably. There were small villages in the area by around this time, with evidence of fishing and cattle husbandry. Although archaeologists have yet to find direct signs of cultivation, it seems likely that human activities as well as natural causes drove the forests' retreat.

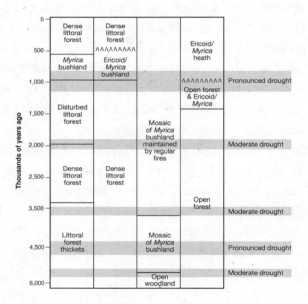

Drier periods (shaded), fire peaks (∧∧∧∧∧∧) and vegetation changes in the southeast over the last 6,000 years (adapted from Virah-Sawmy et al. 2010)

The earliest evidence of cattle in Madagascar comes from bones tossed into tenth-century rubbish pits excavated in the far south, and the south is cattle country *par excellence* today. It should be a good region to investigate the role of introduced livestock in landscape change, although in practice this has been little studied anywhere in Madagascar. The interacting effects of rainfall, soil and grazing pressure on vegetation around a great lake in the southwest have been examined in a recent study, however. The vegetation there is a mix of dry and spiny forest. Pastoralism is not an easy occupation anywhere in the south. The dry season is long, and rains are unpredictable and fail completely in some years. Cattle need to drink, their daily range constricts as free-standing water dries up, and many herbaceous plants in the forest are of poor value as forage. Added to this, the national park established around the lake some years ago is formally off-limits for livestock, and has certainly disrupted the annual cycle of herd movements.

Local livestock owners faced with these challenges drive their cattle herds east of the lake across a high limestone plateau to extensive grasslands beyond, where they spend much of the wet season grazing. Toward the season's end, the limited water sources begin to dry up and cattle are moved back west to browse on woody forest plants. Although the forage is poor, a few pools of water persist in this area, even at the height of the dry season. Goats, on the other hand, are kept year-round on the sandy soils of the lake's western, coastal flank.

Against this backdrop of complex husbandry strategies, researchers examined the number and diversity of plant species in 66 plots established west of the lake on different soils and subjected to varying levels of grazing and browsing. They found significantly fewer plant species and fewer woody plants in sandy coastal plots exposed to intense grazing and browsing. Yet these differences were not found in vegetation on limestone substrates that were also intensely grazed and browsed. Does substrate affect impact? More work is needed to know. Grazing and browsing in and of themselves are not the only issue, moreover. Annual fires set by herders to give cattle a 'green bite' kill some trees and eat away at forest edges. Yet the most resilient species persist and today's swathes of forest in the south, where cattle have browsed and grazed longest, suggest that their long-term impact has been limited.

The journal of a shipwrecked English sailor offers an intriguing historical perspective. Robert Drury was aboard the ship *Degrave* when it broke up on a reef somewhere along the south coast of Madagascar in 1703. He made it ashore and spent the next 13 years as a quasi-slave before escaping, returning safely to London, and publishing his journal in collaboration with Daniel Defoe. Some have questioned its authenticity, but I share Alison Jolly's view: '. . . anyone who has been in the south believes that Drury was there – without necessarily believing every word he wrote . . .'.

Drury's journal is a mine of information about southern landscapes, people and cattle in the early eighteenth century. It evokes the cultural importance people attached to cattle, and detailed descriptions of particular places and forests, woodlands and grasslands are often precise enough to locate on a modern map. In many instances 'the landscape does not seem all that different from what exists today'. Most of the open areas described predate any known human settlements nearby, and the grasslands were likely part of a natural landscape rather than creations of people and their livestock. In other parts of the world, introduced livestock have had rapid and disastrous consequences for endemic vegetation, dramatically shifting the composition and structure of plant communities unaccustomed to their assault, spreading introduced plant species, and sometimes setting the stage for more frequent and intense fires. Perhaps Madagascar's community of endemic grazers and browsers prepared plants for the advent of livestock.

Feral cattle were once common in Madagascar and still roam forests today. Feral animals are not truly wild, but live and reproduce in a wild state after escaping captivity or domestication. Cattle are good at this, and Drury noted frequent glimpses of feral cattle – along with vivid accounts of hunting them. My husband Bob speculated that, by adding to the head count of managed cattle, the feral population extended the human footprint far beyond areas where people pastured their herds. This long struck me as an interesting idea and still does, though I am less convinced now that he was right.

Among all the early written accounts by Europeans, Drury's holds a special place for me. Cattle played a central if elusive role in my life at Hazofotsy. People left the village for a few months each year to tend their rice fields a day's walk away, but cattle were what they talked about and valued most. Some families owned herds of several hundred animals, I learned, though it would have been unthinkably rude to inquire about precise numbers. It struck me as odd that there were so few around the village. Cows with calves were taken out to feed in the forest each day and penned in the village overnight. They were milked, and the milk was fermented into yoghurt. After milking time, the bravest small boys would jostle calves aside to drink from the udder themselves. That was all I actually saw of cattle in this cattle-centric village at the start and end of my days watching *sifaka*. Mostly, the cattle were out in the forest by themselves.

A villager by the name of Tohombinta adopted me informally as a kind of daughter during my time there. One evening, he said he was going up into the hills the next day to look for his cattle and asked if I would like to go with him and his young son. I would, of course. I had had my eye on the distant hilltops for a long time, wondering what grew there and what could be seen from them. We set off early, and walked for an hour or so through forest before the narrow path began to rise. There we paused to drink from a hollow carved in the trunk of a baobab tree. Trees became less common as we climbed, and the hilltop on which we finally flopped down in the late morning was littered with boulders, grasses, little bonsai-like plants and a profusion of wild flowers. It was a different world. Below us, a carpet of spiny forest stretched as far as we could see. I scanned the treetops with my binoculars searching for *sifaka*. Father and son scanned the forest searching for cattle. There were no *sifaka* to be seen and, as far as I could tell, no cattle either. Though we did not find what we were looking for, it was a fine day. I walked back down to Hazofotsy wondering if some of Tohombinta's cattle had just kept wandering further away, heading for a feral life.

Scanning the forest for cattle from hills south of Hazofotsy, 1971
(photograph by author)

What does all this tell us? Both local climate fluctuations and human activities brought about changes in the vegetation of the south, but climate was the main driver of events. In the southwest a thousand years ago, forest cover did not change discernibly in the wake of human settlement. People lived by fishing and animal husbandry rather than clearing land for agriculture, and grazing and browsing by the livestock they introduced had little impact on the landscape. In the southeast at around the same time, forest gave way to grassland during a period of drought, but there is also evidence of small villages in the area by then, and people likely played a part. The question in my mind when I held that hippo jaw in my hands at Taolambiby is answered: the drying southwestern climate caused a hippo disaster. But other regions of Madagascar remained green and wet, and what happened in the southwest does not explain the island-wide extinction of hippos. As one question is answered, new ones open up.

*

The central highlands are only a day's drive from the south, although their high, mountainous terrain and cool climate make them seem much further away than that. A chilly mix of heathland and grassland dominated the landscape for several thousand years after the last global glacial episode until, under warming conditions, heaths receded, grasses spread, and patches of woodland sprang up and gradually expanded into a mosaic of forest, woodland and grassland. A dense layer of charcoal in sediment cores signals a fire peak during the fifth century without evidence of an accompanying oscillation in climate. Could it be the result of human activities? Probably not, because extensive archaeological surveys have found no trace of people in the central highlands at that time. For now, the fire peak's cause remains unexplained.

The first signs of people come from the eleventh century. Farmers living in small, scattered settlements grew crops and herded cattle, and ironworkers also moved into this region rich in iron ore. New, introduced crop weeds show up in the pollen record, suggesting that the mosaic of trees and grasses encountered by early settlers was giving way to fields and more open landscapes. Change came slowly at first. Written reports by European visitors in the eighteenth century still described a landscape of forests, woodlands and grasslands, and so did *The History of Kings*, a collection of oral traditions written down by a French priest in the mid-nineteenth century.

The rise of the Merina Kingdom in the early eighteenth century brought fleeting prosperity to the region. But people steadily used up trees for food, fuel and construction, and small-scale agriculture on poor soils could not feed the growing population. The prospect of famine posed both a threat and an opportunity for the Merina king at the time, the memorably named King Andrianampoinimerina. He exhorted his subjects to redouble their efforts to clear land for agriculture, and by the late eighteenth century vast irrigation works produced enough rice to feed the local population and generate a surplus for trade. Meanwhile, paddy fields encroached further and further into the natural mosaic of vegetation that existed a century earlier. As the human population kept growing, so did demand for wood to cook and keep warm, construct homes and storage huts, or convert to charcoal to fuel a burgeoning iron industry.

Charcoal production for ironworking may have played an important role in the historical reduction of forest cover in the central highlands and beyond. Cultivators need tools, and the best tools are made of iron. The reality of this was brought home to me at Bezà Mahafaly in the 1990s. There were no local ironsmiths at the time, and buying or repairing a plough meant a three-day trip by oxen cart and bus to the regional capital of Toliara. Families shared the few ploughs available, but much of the preparation of soil for planting had to be done with spades, and people were eager for a better option. We loaned money to the mayor to help him attract an ironsmith to the area. The initiative collapsed when the mayor used the money to speculate in the burgeoning sapphire-mining business. The world has moved on since then, but the experience left me with a new understanding: access to iron matters a lot to most people in the countryside, if not to a mayor with other things on his mind.

Artisanal ironworking requires charcoal. Heating iron to hammer into tools takes a lot of charcoal, and smelting iron demands twenty times more: a furnace must be fired to very high temperatures to liberate the metal from its rocky matrix. From the ninth century on, ironworkers set up shop almost anywhere in Madagascar where there were potential customers, and most settlement sites contain slag, the debris from smelting. Today, charcoal production for household use consumes trees at a devastating pace in some parts of Madagascar. A thousand years ago, it may often have been the demands of ironworking that sent people into the forest with axes.

Cattle likely had an impact of a different kind. Historical accounts describe great herds seized as booty during eighteenth- and nineteenth-century Merina military campaigns against neighbouring chiefdoms. Concentrated in the central highlands before being taken down to the coast for export, they grazed and trampled forests and woodlands, and forest edges receded as people fired surrounding areas to improve pasture.

All these activities took a mounting toll on the landscape in the central highlands during the nineteenth century, and the imposition of colonial rule in 1896 brought new pressures thereafter. Stepping back to compare the history of the central highlands with the south, striking contrasts are evident. The climate and vegetation of each region were very different, local fluctuations in temperature and rainfall had their own distinctive timing and character, and the nature and tempo of human environmental impact were different as well. Particularly over the last three centuries, forests receded in the central highlands as the Merina Kingdom used up timber and cleared the land for agriculture. Written accounts by European visitors during this period have no equivalent in other regions of the island, and perhaps distort our perceptions. The Merina political organisation that supported development of commercial rice production was a singular historical event, but some of the changes so vividly described by visitors were probably not unique to the central highlands. The earliest big towns in Madagascar were in the northwest. It seems reasonable to expect parallels in what happened there.

*

Even though only a day's drive away from the central highlands, the northwest has its own distinctive climate history. Topography and the complex workings of the global climate system drive differences between the two. Each is linked to a separate regional climate system in Africa, with shifts in the northwest bearing a strong resemblance to events in east and north Africa while those in the central highlands echo southeast Africa. Quite why this is so remains to be determined, but it is certainly one more reason why island-wide scenarios for the past do not work.

Near the village of Anjohibe, some 70 km northeast of the modern city of Mahajanga, lies a huge cave where stalagmites stand sentinel along its 5 km or so of passages. Five stalagmites from Anjohibe and Anjokipoty, 16 km away, provide the fullest environmental record of the past after the last glacial episode for any region of the island over the last 10,000 years. You can almost experience the weather! Pollen from a sediment core at Lake Mitsinjo 115 km away adds geographical spread to the picture, and archaeology another dimension still.

The stalagmites reveal major changes. For most of the period they span, vegetation 'behaved' as one would predict under conditions controlled by climate. At first the climate was wet, wetter than in any more recent period, and the landscape tree-covered. After this, wet and dry conditions alternated in rapid succession, sometimes within a decade, and trees and C4 grasses spread and flourished by turn. Two major droughts followed, and the climate

Peterson Faina beside a stalagmite in Anjohibe Cave
(photograph by Laurie Godfrey)

became generally drier in their wake. Now the vegetation was a mix of C3 and C4 plants, perhaps a palm-studded grassland like that covering parts of the northwest today.

The tightly linked climate and vegetation signals diverge slightly for the first time around 2,500 years ago: signals of grassland are stronger than they 'ought' to be. We know there were foragers in Madagascar by then. Were there farmers too, clearing northwestern forest for fields? It is possible, although there is no archaeological evidence for this, and the decoupling of signals is more likely a wobble arising from the complex techniques involved in the analysis.

What happened closer to the present was certainly no wobble. The fifth century started out with drying conditions and an increase in C4 grasses – just what you would expect if climate was the main driver – but then a dramatic decoupling of signals followed. Around the eighth century, a massive and abrupt shift to C4 grasses took place with little or no change in climate. Estimates of when exactly this happened vary, but the shift itself is clear in every stalagmite examined.

Set against these findings, the sediment core from Lake Mitsinjo 115 km away tells a different story – of drought, not stable climate conditions. The lake, formerly deep and surrounded by forest, dried up and filled again only when good rains returned in the eleventh century. This could be an instance of extreme local climate variability, or reflect the challenge of dating sediment cores accurately. Changes were indeed sometimes very local, stalagmite analyses tell us: while grasslands surrounded Anjohibe between the eleventh and eighteenth centuries, trees were springing up around Anjokipoty just a few kilometres away. For now, the discrepancy between the record from Mitsinjo and the stalagmites serves mainly as a reminder of how much there is still to learn.

The most plausible interpretation of the evidence overall is that people began clearing land in the northwest from the eighth century, the impact of their activities perhaps amplified by local climate events in some places. If this is correct, it is puzzling that archaeological surveys have yet to find traces of human settlement. But the surveys so far are quite few. My prediction is that future work will uncover vestiges of villages, and that by the eighth century or earlier people were indeed living in the area, establishing fields and converting woodland to grassland for pasture.

Edging forward in time, the twelfth century saw the appearance of towns in the northwest, including the great trading port of Mahilaka 80 km or so up the coast from Anjohibe. Cultivating crops in the countryside would become more than a matter of feeding your own household. There were other mouths to feed now, as urban markets for food developed. The pace of forest clearance in the hinterland of Mahilaka likely accelerated, and perhaps the impact of Mahilaka's development reverberated as far away as Anjohibe. But change could be short-lived. Mahilaka's eventual decline was associated with the abandonment of most surrounding villages, signs of burning diminish, and tree pollen becomes frequent again in the record. It seems likely that forests returned, at least partially, as human pressure decreased.

The northwest adds a third dimension to our picture of how climate and human activities drove changes in Madagascar's vegetation. Wetter conditions became somewhat drier there over the last 10,000 years, though that simple statement overlooks local variation between nearby places and also ignores a lot of blips through time. Although blips in the grand scheme of things, they would certainly have presented challenging weather for plants, animals and settlers too, when they began clearing land for fields between the fifth and tenth centuries – considerably earlier than in the central highlands, and long before discernible human impact in the south. The collapse of Mahilaka and subsequent regeneration of forest in the area warns against simple, unidirectional scenarios, however. For the last few centuries, the record is thin. In the absence of coordinated, state-level organisation driving events as it did in the central highlands, however, it is unlikely the northwest experienced an equally profound transformation.

*

A simple story starring *Homo sapiens* as villain does not add up. Vegetation changes due to human activities are noticeably missing anywhere on the island for thousands of years after the first signs of people appear. They become clearly evident little more than 1,000 years ago, but only in the northwest. With population growth and expanding economic activity over the next few centuries, one might expect rapid, island-wide conversion of forest to fields and pasture. But this is not the picture that emerges. Climate fluctuations drove some changes in vegetation, people drove others, and a combination of both drove many. Just as there was no single moment of human settlement in Madagascar, there was no single moment of environmental disaster. As a demonstration of instant catastrophe triggered by people, the record is a distinct failure.

People made their living differently depending on where they settled, and the timing and drivers of change varied too. Reflecting on discoveries at the rock shelter at Lakaton'i Anja, Bob and the Anja team wrote: 'The activities of foraging populations have environmental consequences that differ in both degree and nature from those of Iron Age farmers and pastoralists, and changes in palaeoenvironmental proxies interpreted as signalling 'human arrival' may in fact be signals of a change in human economy'. Laurie Godfrey and her colleagues have recently developed this idea into a 'subsistence shift hypothesis'. Initially, reductions in forest cover depended not on the simple presence of people but on what they did for a living, and people made their living differently depending on where they settled and the skills and knowledge with which they arrived. This helps explain why historical changes in vegetation often seem to have been least where people settled earliest, with some of the most intact plant communities in coastal areas.

The subsistence shift hypothesis provides a compelling way to think about the history of human impact on forests until the last few centuries. In the central highlands certainly, and perhaps in other regions too, state-led activities became increasingly important and may have done more to clear forest than subsistence farmers ever did. It would help explain why the mosaic of forest, woodlands and grasslands early settlers encountered in the central highlands ended up becoming an expanse of grassland in many places. Perhaps an 'organisational shift hypothesis' is needed as well, extending beyond change in subsistence activities to broad societal transformation. In addition, the history still contains plenty of gaps and puzzles. Undoubtedly the biggest gap is for the eastern rainforests, with virtually no evidence until the twentieth century. Forests linking the central highlands to the coast remained largely intact well into the nineteenth century according to historical accounts, but the reliability of these reports is uncertain and there are vast areas about which nothing is known at all.

A thousand years ago, in the forests, woodlands and grasslands of Madagascar roamed animals now extinct. They were survivors of whatever climate changes had thrown at them and their habitats over millions of years. Did they die out because too many of these habitats disappeared as a combined consequence of local climate shifts and human activities, or did people play a more direct role in their demise?

*Elephant bird imagined (artwork by Wallace Blanchard,
from* Pearson's Magazine, *1905)*

CHAPTER 9

Disappearing Giants

A group of elderly villagers on the west coast fared better with hippo encounters than I did at Taolambiby, at least according to one account. While I merely held a sub-fossil, they told visiting researchers of actually seeing or hearing a creature they called *kilopilopitsofy* in 1976! One man gave an imitation of the *kilopilopitsofy's* call, sounding uncannily like a living hippo in Africa – yet he said he had never seen or heard an African hippo. The animal they described closely matched a written account four centuries earlier of cow-sized, hornless animals in the hinterlands of modern Taolagnaro on the southeast tip of the island. Although the writer never saw these animals himself, he faithfully recorded what others told him: they had dark skin, a large mouth, huge teeth and big, floppy ears. His description is now generally thought to be of a dwarf hippopotamus.

Researchers report such conversations gingerly. Perhaps the elderly villagers were quietly having a good laugh at their visitors' expense. But their accounts are not to be ignored, despite

195

the lack of physical evidence to back them up. It is at least possible that the last hippopotamus was grunting within our lifetime. Madagascar was certainly a fine place to escape naturally fluctuating conditions and avoid contact with people after their arrival. Its islands-within-an-island character offered animals many opportunities. When climate conditions deteriorated or people appeared in one area, there was a good chance of finding more welcoming circumstances in another. This meant that widespread species could continue to flourish in one region while dying out in another, and animals capable of long-distance travel could even up sticks and move. Hippopotamuses and a large number of smaller mammal species lived on further north after disappearing from the southwest, and descendants of many of the small mammals can still be found to this day. But not the hippopotamus.

Escape did not always work. Madagascar's endemic giant tortoises and elephant birds, large-bodied lemurs and hippopotamuses all disappear from the sub-fossil record within the last thousand years. There are very few well-dated bones after the tenth century, and none fall within the last 500 years. Sightings like those of the villagers in the west are the only hint that any of these animals persisted closer to the present. The impact of early settlers on the vegetation around their villages does not explain island-wide extinctions. Forests, woodlands and grasslands persisted that between them could have harboured these animals. What, then, brought about their demise?

*

Madagascar's big animals certainly did not disappear overnight after people arrived around 10,000 years ago. They were still alive and well 4,000 years ago, in fact, at the time of the earliest foragers' camp site discovered in a rock shelter in the far north.

For a long time, humans and wildlife shared the land without discernible consequences. A broad chronology of what happened thereafter comes from sub-fossil remains with reliable radio-carbon dates, which inject time where before there was little or none. The dating is a monument to technical advances in science – and hard work. I got a firsthand glimpse of both in 2016 at the fiftieth anniversary celebrations of the Duke Lemur Center in North Carolina, where a poster session was held outdoors on a warm autumnal day. Thirty or so researchers stood by their easels describing all manner of research on lemurs, and I intended to visit each one. My first stop was at a poster about the value of captive animals for isotope studies, the work of a young woman who introduced herself as Brooke Crowley.

Brooke's name was well known to me. Over the next twenty minutes she gave me a tour of the horizons of isotope research, at which point the skies opened and everyone fled for cover. Hers turned out to be the first and last poster I saw that after-noon. If I could only see one, I am glad it was that one. Brooke's work established the chronology of mega-faunal decline, and helped ferret out clues to the diets of extinct species. Meeting the person behind important work is always interesting and, on this occasion, I received a great tutorial as well. In the coming years, the rising generation of researchers outside Madagascar as well as within will surely fill in many pieces of the puzzle still missing.

Some of the bones and bone fragments Brooke studied came from museum drawers, others from recent field expeditions. Together they represented a wide array of species from many places around the island, with the notable exception of the east. Bone preserves better under dry conditions, and palaeon-tologists have been less successful finding bones in the wet east (and have probably spent less time looking, too). Brooke grouped bones by species, estimated body size, estimated age of the remains, and the region they came from. The first hints of decline appear in the seventh century, indicated by fewer bones from extinct species, and they all but disappear from the record over the next several hundred years. Fewer than 10 bones date to within the last thousand years, and the 'youngest' is from the fifteenth century.

Brooke, Laurie Godfrey (the Bezà Mahafaly grave digger!) and their colleagues have recently zeroed in on 195 sub-fossil bones and bone fragments all dated within a 1,500-year period, between 2,000 and 500 years ago. This spans the period before and during intensifying human settlement, population increase, and shifts in the way people lived. Their study compared the ratio of extinct to still-living species at hundred-year intervals, making it possible to probe the pace of decline in addition to its broad timing. A sudden drop-off in the bones of now-extinct species compared to living ones would suggest a precipitous decline, while a more gradual reduction would point to a long, slow dwindling. A clear pattern emerged from the analysis. Bones from extinct species are dominant until the seventh century, but then they sharply decline in relative number and by the eleventh century they are rare across the whole island, with most bones in the sample belonging to species still living. A different kind of clue flags up these centuries as well. It comes from fungi (*Sporormiella* spp.) that feed on dung, signalling the presence of large-bodied plant-eating animals. A reported drop in the abundance of *Sporormiella* in Madagascar starting in the seventh century may be further evidence of the decline of the mega-fauna, but the significance of such fungal changes is debated and I prefer to stick with bones.

Chance surely played a role in producing the bone chronology

– which bones happened to be preserved, and happened to be found and picked up by researchers – but the results are very consistent. Some animals, like the *kilopilopitsofy*, may have lived on in remote areas much closer to the present, but that did not change the ultimate outcome. The bone chronology points clearly to the period when the human population was growing and people began settling in villages as one of rapid loss for the mega-fauna. But correlation alone does not add up to an explanation for the extinctions. If indeed people were responsible, what were they *doing* that they had *not* been doing further in the past?

*

Giant lemurs and tortoises seem obvious targets for the cooking pot, sitting ducks for even a moderately competent hunter. Most giant lemurs clambered around slowly in the trees. Giant tortoises moved at, well, a tortoise pace and were even more poorly equipped to fend off attacks from new predators in human form. If the decline of these animals accelerated as the chronology indicates and if hunting were the cause, one would expect settlement rubbish piles and pits to be filled with remains of their bones, left over from meals eaten long ago. But they are not.

The remains of extinct animals are rare or non-existent in settlement sites anywhere on the island. This could be because people butchered animals on the spot after killing them and only brought back to the village chunks of meat that could be carried easily, or because wild carnivores and dogs made off with bones left over from long-ago meals around the hearth. The virtual absence of remains is nevertheless perplexing and contrasts sharply with New Zealand, another big island, settled even more recently than Madagascar. There, early settlement sites contain thousands of bones of now-extinct, flightless birds called moas.

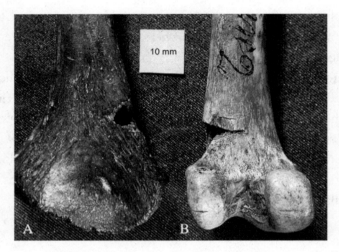

Cut- and chop-marks made on bone during butchery
(Perez et al. 2005)

Except for the foraging camp at Lakaton'i Anja in the far north, evidence for hunting in Madagascar comes from cut- and chop-marks on bones – with no sign of the tools that made the marks or the people who wielded them. Most of the bones are from large caches discovered in the southwest. How these caches formed is not known, nor is it clear how many individual animals are represented by the jumbled bones. Perhaps they are vestiges of butchery camps, or perhaps floods washed bones downstream from butchery sites further away to pools where they were trapped and slowly accumulated. But the caches are treasure troves of evidence, regardless of how they came about.

Sorting carefully through many hundred bones from the southwest, Laurie, Brooke and their colleagues identified 24 cut- or chop-marked bones with reliable dates falling between the first and fourteenth centuries. Thirteen of the 24 bones belonged to five now-extinct species: two were large-bodied lemurs – the giant ruffed lemur (*Pachylemur insignis*) and monkey lemur (*Archaeolemur majori*); two were elephant birds, *Aepyornis maximus* and *Mullerornis sp*; and one was a hippopotamus, *Hippopotamus lemerlei*. The other 11 bones belonged to *sifaka*. These all came from Taolambiby, where *sifaka* still live in the forest today. The Taolambiby collection also contained many ring-tailed lemur bones, but none show signs of butchery. This seems odd. I wonder if people did not hunt these lemurs for some reason, or perhaps butchered them in a way that left no mark. The estimated ages of butchered bones suggest that the biggest animals were mostly hunted in the ninth century. The *sifaka* bones, in contrast, date to the tenth to twelfth centuries.

This is the best evidence so far that early settlers hunted, and it is singularly underwhelming – a handful of cut- or chop-marked bones discovered amid hundreds of bones showing no evidence of butchery at all. The marked bones come from one or a very few individuals of extinct species and a slightly larger number of individual *sifaka*. Interpreting even the latter as evidence of a slaughter is a stretch, particularly since the unknown number of *sifaka* to which the bones belonged may actually have been hunted over a period of years or even decades. The dating is not fine-grained enough to tell. The genetic make-up of modern *sifaka* in the southwest gives no indication of a bottleneck during the last two thousand years, moreover, which would signal a population collapse in the face of heavy hunting. Maybe people turned to smaller *sifaka* as larger prey became scarce and eventually disappeared, and then hunting traditions gave way to taboos that protect some surviving lemur species, including *sifaka*, to this day.

Bones shed light on the importance – or not – of hunting pressure along a pathway of decline. Animals occupy a place in the human imagination far beyond their value as food, however, and bones do not speak to this. Cave paintings in many regions of the world tell us that humans have long attached symbolic and cultural importance to animals, especially big ones, and to hunting. What about Madagascar? A new discovery of prehistoric cave art hints that people viewed the big animals alive in early times as special even as, paradoxically, they hunted them. The discovery was made in the interconnected, narrow passages of Andriamamelo Cave in the west, one of many limestone caves in the region. Until then, the only well-described cave art in Madagascar was in the southwest, with images depicting symbols and geometric designs. Andriamamelo's art is different.

Cave painting of a hunt (Burney et al. 2020)

Covering an area 3 metres long and 1 metre high on two flat, pale grey limestone panels, the art depicts clusters of humans, animals and strange animal-human hybrids, along with a scattering of enigmatic letters and symbols. One of the scenes appears to be a hunt, where a person brandishes a weapon of some sort and an animal lies on its back with its legs in the air. 'This animal's profile is consistent with that of the extinct sloth lemurs', the authors note cautiously. Sub-fossil remains of sloth lemurs are common in the region, although none show evidence of butchery. When did the artists who produced these images live? Efforts to extract dates from the charcoal they used were unsuccessful, but

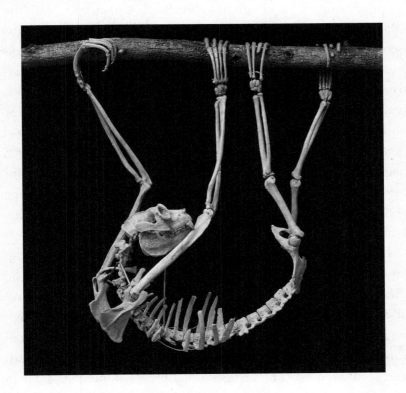

Skeleton of a sloth lemur, Palaeopropithecus kelyus *(Duke Lemur Center Museum of Natural History / David Haring)*

the letters and symbols suggest possible ancient connections to northeast Africa, Arabia and Borneo. Cattle do not figure among the animals depicted despite their longstanding cultural salience for many Malagasy people, and this too hints at a period before the time of their introduction in the tenth century. For now, however, Andriamamelo is an intriguing place of ghosts – and many unanswered questions.

*

From bones and art to . . . eggs. Bony evidence that people hunted elephant birds is just as sparse as it is for giant lemurs, hippopotamuses and giant tortoises. Cut-marks on the 10,000-year-old bones of an elephant bird have become a new benchmark for the earliest human presence on the island; marks have also been found on the leg bones of three individuals with somewhat younger dates, and on two bones dated to the eighth century. These are scarcely indications of a hunting *blitzkrieg*, and settlement sites contain no remains of roast leftovers and not a single clearly cut-marked bone. This is perhaps surprising, because ostrich meat is fine fare and one supposes elephant birds would have tasted good too. What *has* been found in huge abundance is elephant bird eggshell. In fact, it is the most easily found trace of any of the mega-fauna.

Dense concentrations of broken eggshell on southern and far northern beaches are probably the remains of colony nesting sites where birds converged from inland to reproduce, taking advantage of the warm sand to incubate their eggs. But eggshell fragments picked up from the surface of the ground or exca-vated by archaeologists in camps and settlements in the south and southwest are not vestiges of nesting colonies. At least in those areas, people were harvesting broken eggshell or stealing whole eggs. Piles of elephant bird eggshell in rock shelters at

Velondriake in the southwest date from 2,000 years ago all the way up to the fourteenth century, and shell fragments have been found at 50 or so settlement sites in the south from the ninth century on, although only in large quantities at a handful of these sites.

What would people have collected eggs for? Their contents would have made an unimaginably large omelette. With a volume of 11 litres, a single egg laid by the largest of the elephant birds would have been equivalent to 180–240 chicken eggs. Even if such culinary *tours de force* were actually created, they would quickly have disappeared without trace into a large number of stomachs. Emptied whole eggs would have made compendious liquid containers, and there is a thin thread of support for this use: people transported rum to and from market this way according to two nineteenth-century accounts. But it would take a lot of rum transportation to drive elephant birds to extinction.

Eggshell was perhaps attractive for ornamental purposes. Ostrich eggshell beads are common in archaeological sites in Africa, and beads were important exchange items in southern and eastern Africa for many thousand years. Early foragers arriving in Madagascar from Africa may have brought with them a liking for shell beads, though the only beads fashioned from elephant bird eggshell (and seashell) found so far come from eighth-century deposits at Velondriake. The deposits also contain eggshell fragments with thin, deep incisions, and the edges of some seemed intentionally chipped and scraped. The purpose of these modifications – if that is what they are – is unclear, and Velondriake is still the only place where anything like this has been seen.

Whatever people were using whole eggs or eggshell for, they were using a lot, and the theft of eggs may have played a bigger part than hunting in the decline and extinction of elephant birds. Preliminary dates indicate that fresh eggshell was the likely source of the Velondriake beads, and dated shell fragments from Talaky in the far south also seem to have come from eggs laid by birds living during the time the site was occupied. The importance of egg theft in the extinctions still hangs in the balance. Maybe people cracked open fresh eggs, but they could equally have harvested fragments after eggs hatched naturally. There is as yet no way to distinguish between cracking open and hatching and, thus, between a dead and living chick.

Feathers have it over beads when it comes to adornment in my view. Plumes of the elephant bird's tail would have far outshone the ostrich feathers bedecking fancy hats today. A thirteenth-century Arab text referred to feathers 'a size greater than a span and a half' of a monstrous bird, the *roc*, gathered for export to Aden and eventual trade. Picking up this tale, Marco Polo wrote a fantastical account of a bird in Madagascar big enough to pick up an elephant in its talons, and of messengers from the 'grand khan' sent to bring back its feathers. The story is problematic for many reasons, not least the absence of elephants in Madagascar, and in fact it was rubbished within two hundred years of being written. In our own times, it has been pointed out that the Malagasy word for feather and bamboo is the same, *volo*, and that the returning messengers almost certainly presented his majesty with a giant bamboo stem for water storage, not a feather.

These bizarre stories are weakly grounded in history, to put it mildly, but they may nevertheless be illuminating. Traders setting up outposts in the south would surely have been on the lookout for interesting and exotic items for export, and I wonder if beads fashioned from elephant bird eggshell, the eggs themselves or even plumes will surface some day in Africa or Arabia. The feathers must have been magnificent. Had I been an early trader along Madagascar's coastline I would have been chasing elephant birds for those plumes. Then again, perhaps I would not. No fewer than three of the six species listed as 'the most dangerous birds in the world' by the *Encyclopedia Britannica* are living relatives of elephant birds! Here is an entry in the journal of a Frenchman called Ruelle in 1667, describing his encounter with an elephant bird:

> 'We met a terrible winged dragon . . . one day out hunting with a soldier who was accompanying me, we saw one at the foot of a tree, that rose into the air as soon as we saw it, with a horrible hissing and red eyes of fire. When he had risen to the height of two spears he began to fall upon us, so I fired a shot that caused him to fall at my feet, with a frightful writhing about. My soldier companion killed it with another shot. It was 15 feet long. The head and body big like a calf tapering from the wings to the tail. Its scales were black and yellow; the wings were three feet in diameter. It had two rough and scaly feet. The body, we skinned to present it to Mr. Marquis de Mondevergue'.

Some might set Ruelle's account aside as unreliable, but if their living relatives are anything to go by, elephant birds along with crocodiles were by far the most daunting beasts among Madagascar's generally amiable wildlife. Consider living cassowaries. Weighing about 60 kg, they are smaller than the biggest elephant birds were, but with a uniquely long, razor-sharp nail on each foot, a reputation for ferocity, and a 'malignant eye'. A blood-chilling report of a cassowary attack in New Guinea in the 1980s leaves little doubt that this reputation is deserved:

a man bent on stealing eggs approached a nest where a male cassowary was on guard, and the cassowary attacked; the would-be thief managed to cut off one of its legs with a machete before it leapt one-legged into the air and disembowelled him with a swipe of the dagger-like claw on its remaining foot. With his intestines stuffed back into his stomach, the man survived the encounter. The cassowary did not. Whether each got what he deserved is questionable, but together with Ruelle's tale the episode does suggest that getting one's hands on fresh elephant bird meat, eggs or plumes could have been a dangerous business.

*

Madagascar is huge, and its human population was relatively small until the late twentieth century. Population density is low even today compared to many regions of the world. No island-wide catastrophic change in climate has visited Madagascar in the last thousand years. There were plenty of remote places to which animals could escape from people, and no wave of hunting or egg destruction overtook the island's mega-fauna in the wake of their arrival. People hunted from time to time and may have stolen eggs, but these activities look more opportunistic than a way of life. They seem desultory, insufficient to bring about extinctions. North America, where there is much more evidence, suggests otherwise. Compounding effects can be fatal.

More than half the large mammal species in North America around 12,000 years ago disappeared within a period of roughly 1,500 years. Could hunting pressure alone account for this? A computer simulation of human settlers and 41 large, herbivorous mammal species they hunted – 30 of these species now extinct – suggests that it could, easily, and offers insight into the trajectory of extinctions in Madagascar.

The simulation considered different variants of many factors. For people, these factors included estimates of the date of the first substantial human presence on the continent, the rate at which the population grew, and the level of individual hunting skills; for prey species, the factors included geographical range, body size, and population density and growth rates. Results of the simulation are clear. Almost regardless of the values used for each factor, human population growth and hunting led to mass extinction. Even really incompetent hunters achieved this outcome, and it actually proved hard to simulate conditions under which all species survived. Large-bodied species were more likely to go extinct than small ones according to the simulation – as indeed was the case. Scenarios that best predicted the extinctions assumed only modest human population growth rates and hunting ability. Predictions of the time that elapsed between initial decline and extinction ranged between 801 and 1,640 years – the blink of an eye in Earth history, dozens of generations in human history.

North America is half a world away from Madagascar and simulations are not real life. Yet the finding that modest rates of hunting would likely eliminate large-bodied, slowly reproducing species across a huge continent over a period of less than 1,000 to a little over 1,500 years suggests the path to extinction in Madagascar too. Giant lemurs and tortoises matured and reproduced slowly, and lemurs probably lived at low densities in large ranges. (Big tortoise concentrations, on the other hand, would have been easy targets.) The bone chronology of decline shows numbers of animals belonging to the largest-bodied species starting to dip in the seventh century, and from the eighth century archaeology signals a growing human population with the arrival of new settlers. Even occasional and incompetent hunting combined with the biological vulnerability of large animals was probably enough to tip the scales. Extinction by a thousand small cuts?

Some evidence is difficult to reconcile with a scenario of elimination by opportunistic hunting alone, with a fatal tipping point as the settler population grew. Climate as well as human-driven reductions in forest cover and increases in open habitat likely compounded the impact of hunting in a synergy of effects that differed regionally depending on how people made a living. In the southwest, major drought coincided with and likely drove the decline of large animals. In the northwest, the Anjohibe Cave stalagmites signal a clear shift to more open country in the absence of climate change around the eighth century. Dating the sub-fossil bones of now-extinct species excavated there proved difficult, and the few reliable dates – from monkey lemurs, elephant birds and hippos – all fall before the sixth century. This puts the 'last sighting' of these species several hundred years *before* the shift to grasslands and well before the first archaeological evidence of human settlement. The sample size is small, and perhaps younger sub-fossil bones were excluded by chance, but this seems unlikely. There are bones from introduced and endemic species found in the region today that date to the past few hundred years, and the bones of extinct species should be among them if they were still living then.

For now, the jury remains out on the precise timing and causes of the extinctions around Anjohibe, the most comprehensively studied and dated of any site in Madagascar, and far less is known about the sequence of events in other regions. The clues all point to a complex web of effects across the island's landscapes, however. Low or moderate levels of hunting, forest clearance for agriculture, timber, firewood and charcoal, livestock husbandry and local climate fluctuations likely worked in varying combinations at different times and places to bring about the long decline and eventual disappearance of animals that live on only as ghosts today.

*

The extinctions in North America have been described as an '. . . ecological catastrophe that was too gradual to be perceived by the people who unleashed it'. Something similar could be said about events in Madagascar during the last 1,500 years, with natural fluctuations also contributing in some areas. The colonists' story took a simpler and harsher view a century ago. Instantaneous, wanton destruction was its keynote, not change too gradual to be perceived. Characterising the mere presence of people as a perfect storm, end of story, is not only mistaken but also a disaster for conservation efforts today.

Chopping wood near Hazofotsy (photograph by author)

CHAPTER 10

The Axe's Thunk

It was there every year until it wasn't. Thick forest covering the slopes of a deep valley beside the road would come into sight as the car edged down the steep, rocky escarpment of the Mahafaly plateau toward the flood plain of the Onilahy River in the southwest. I always had an eye open for lemurs in the tree tops as we crawled along, though I never saw one. In 1990 a small, neat cleared patch appeared on the forest edge. Within three years there were no trees left in the valley, just fields of maize. Then the fields disappeared, and the valley became a desolate place of grass and rutted paths made by grazing livestock. It was sad to see. It was stunning too. A forest that had persisted for so long vanished so quickly. Why did this happen?

You are never far from the sound of an axe in Madagascar, and its echoing thunk has been part of the soundtrack of my hours in the forest for fifty years. The buzz of chainsaws makes its own contribution these days, particularly in the east, but that sound is still rare compared to axes. They are not the huge kind wielded by lumberjacks in America. Axes carried by people in the countryside of Madagascar typically have small iron heads hafted on rough-hewn wooden handles. They do not look equal to the task of clearing whole forests, but people use them with skill and precision. Under their assault, big trees topple fast.

Close to half the forest that existed 50 years ago has disappeared since then, more is lost every year, and the accelerating impact of global changes in climate, particularly in southern Madagascar, is likely to drive these losses faster. There is little research on the extent of grassland degradation although that gap is finally starting to be filled, but the trajectory for forests is not in dispute: cleared, hollowed out, fragmented or degraded forests are common features of the landscape. The pressure on forest-living wildlife grows as their habitats disappear, and hunting, international trafficking and competition from introduced species take a further toll. Events play out in a shifting and unstable political setting that sends ripples of lawlessness and despair across the country. In theory, evolving national environmental policies offer encouragement to those bent on protection. In practice and for now, the impact of these policies is limited at best. Madagascar's remaining natural environments are under grave threat.

Ignorant, pyromaniacal peasants with axes and fire were the scourge of the island's forests in the colonists' story. A more benign version has emerged in post-colonial times, with poverty and the need to survive replacing ignorance and wilful destruction as explanations for forest clearance. But too often it remains fundamentally the same story of villagers in the countryside levelling forests. There is a different way of viewing human impact, however, both today and in the past. Madagascar is caught up in complex and enduring struggles for control of the land. Those struggles initially pitted Malagasy communities against one another. At the end of the nineteenth century, the French colonial government entered the fray. The contest for control persists today, with new as well as longstanding players. Individuals and businesses seek profits from natural assets, national and foreign governments seek to promote development, and conservationists seek to prevent further environmental destruction. Rural folk aspiring to new fields in order to put food on the family table are just one group among the contestants, and often on the losing side – along with the environment.

*

I have spent a good bit of my adult life sitting on rosewood stools in our living room, or so it feels. The origin of each stool with its three stubby legs and round carved seat was a single slice cut from a big tree. Rich burgundy in hue, beautiful, heavy, hard and surprisingly comfortable, my affection for them was simple until they came to symbolise an unfolding disaster.

Today, rosewoods are under imminent threat of extinction in Madagascar. The wood is highly sought after from 10 of Madagascar's 48 species of this genus, *Dalbergia*. It is prized precisely for the qualities I loved in our stools. Most exports are shipped to China, where the demand for expensive rosewood furniture is high. In one recent year alone, at least 164,000 logs made their way out of the country illegally, with a value of almost £167 million. Those wielding the axes that felled these trees received about 40 pence for each tree felled. The trade exploits not just rosewoods but also the Malagasy woodsmen who give up their natural inheritance in return for a pittance. Although selective cutting results in lower direct loss of forest than clear-cutting, its long-term indirect effects add up to widespread destruction. Government efforts to prohibit rosewood logging and export have largely failed, and a flourishing trade continues illegally. Attracting anguished headlines alongside ineffectual measures to end it, the rosewood trade may be the best-known

Illegally harvested rosewood logs (Dalbergia spp.)
impounded in the east (AVG/Solofo Ralaimihoatra)

activity depleting forests today, but the maize fields of the south-west remind us that it is by no means the only one.

People were clearing forests far into the past, but many features of Madagascar's modern landscapes emerged only in the last hundred years or so. Creation of the colonial state in 1896 unleashed a cascade of assaults, with a new government hungry to increase revenues and new settlers eager to make their fortunes. Successive colonial administrations offered liberal land grants to individuals and companies from France and neighbouring islands as an enticement to establish cash-crop plantations and exploit the island's forests, cattle and minerals. Colonists assumed control of the most fertile areas in valley bottoms in the east and began growing crops, particularly coffee and rice, to feed the export market. Timber was a favoured export commodity too, graphite mining flourished, and the transport and communications network designed to support

216

the extraction and export of resources made further inroads on the landscape.

As colonists expropriated the best land and its assets to grow export crops, displaced local cultivators moved further into the forest or to higher, less desirable forested slopes on hillsides. They were subsistence farmers, their harvests not for export or even primarily for cash but to put food on the table. They used a system of shifting cultivation. The cycle begins with felling trees and burning undergrowth. The ashes from burnt vegetation release nutrients into the soil, making it a form of agriculture that requires few inputs except labour. It is not surprising that many poor rural households relied on it then and continue to do so. Forest regenerates if fields lie fallow long enough after a period of cultivation, though not if the cycle of clearance-cultivation-fallow-regeneration is cut short, as commonly happens.

Descriptive accounts and colonial logging records leave no doubt that clearance accelerated under colonial rule, particularly in the east and central highlands. Contemporaneous guestimates of the decline varied from 3 to 7 million hectares lost between 1900 and 1940. That is an area somewhere between the size of Belgium at the low end and Ireland at the high. The wide range of figures was a result of differing definitions of forest types (a problem accentuated in Madagascar by the diversity of vegetation), poor or non-existent maps and possibly 'adjustments', conscious or not, to strengthen a particular argument. The loss was substantial even at the low end of the range, and likely much higher in relation to the actual forest cover at the outset. Eventually, pressured by the evidence of destruction, the authorities drafted a forest code that forbade tree cutting on public lands and outlawed burning near forests. Twelve nature reserves were established too, over time, off-limits to people living around them.

The second half of the twentieth century brought the age of aerial photography and satellite images. In principle, tracking vegetation changes accurately was now possible. In practice, doing so turned out to be a 'messy and difficult task', with technical challenges and the same old problem of defining vegetation and forest types clearly. The particular objectives of analysts mattered too, shaping the kinds of questions asked and results highlighted. Interpretations of aerial photographs taken in the 1950s suggested that there were somewhere between 16 and 20 million hectares of forest at that time, depending on what was counted as forest. Satellite images of forest cover around the year 2000 yielded estimates ranging from 9 to 17 million hectares. A new analysis spanning the last 50 years estimates that around 44 per cent of Madagascar's forests have been converted to other uses during these decades. Those left are increasingly fragmented: around half of the remaining forest lies less than 100 metres from the forest edge.

The range of estimates is broad, and island-wide trends remain hard to pin down. One commonly cited study concluded that around 1.7 per cent of forest cover was lost each year during the 1970s and 1980s – a higher loss rate than in the decades before or after. But percentages vary from one study to another. Regional studies, using the same types of satellite data and consistent definitions and methods over time, indicate that forests have been diminishing in most regions, dramatically so in the southwest, but that boundaries have remained stable in some areas or even expanded.

Why does being careful about numbers matter in the face of widespread, undeniable loss? The answer has to do with the insidious power of numbers with false precision, like the oft-repeated statement that 90 per cent of Madagascar's 'original' forests have been lost. That figure is simply a guess, and the date of 'original' never given. There is still no vegetation map of the island when people arrived, and you cannot determine the scale of change when you do not know what existed to be changed. New numbers derived from aerial photographs and satellite images are based on carefully assembled evidence and – for that reason, paradoxically – claim less certainty.

*

Robert Sussman and I met as graduate students in Madagascar in 1970. Bob died too young in 2016. By then he was a professor of Anthropology at Washington University, longtime Madagascar maven, and good friend. His PhD research site at Antserananomby lay a few kilometres north of the Mangoky River, one of the great rivers in the west. Bob was studying how different lemur species divide up forest resources between them, and listening to him talk about Antserananomby made me eager to visit. When we finally set out into the forest together in 1974, I was puzzled that Bob seemed embarrassed as well as happy. I quickly understood why. No researcher is entitled to so many lemurs in one place.

Antserananomby forest was alive with lemurs, like no forest I have seen elsewhere in Madagascar or the rest of the world. The green canopy of *kily* (*Tamarindus indica*) trees over our heads was filled with *sifaka* bouncing from trunk to trunk and brown lemurs (*Lemur fulvus rufus)* running along boughs, tails swaying like pendulums; on the ground, ring-tailed lemurs scampered by. Five nocturnal lemur species came on stage as a different cast of characters after dark. It was a joyful forest, and a wonderful place to explore how animals divided up its larder. Returning in 2004, researchers found the Antserananomby forest gone, replaced by fields of maize. Villagers reported that newcomers who did not respect traditional prohibitions were responsible. Of lemurs and other wildlife there were traces only in surrounding forest fragments.

As forests disappear, so do the animals that depend on them for a living. Between surveys in 2010 and 2015, the percentage of lemur species listed as endangered or critically endangered by the International Union for the Conservation of Nature rose from 31 to 72 per cent. This is partly a result of subdividing populations previously considered as a single species into several new ones. With limited geographical ranges and low numbers, the new species automatically qualified for the highest category of threat. Debate about the justification for some of these subdivisions masks the reality that lemurs are losing habitat and the survival of many species, however defined, hangs in the balance – and so does the long-term survival of plant species that depend on lemurs to pollinate their flowers or disperse their seeds. Lemurs play an important part in these vital functions, and the extinction of giant lemurs has already 'orphaned' some plants. According to a recent study of over 3,000 plant species, the geographical ranges of their main lemur dispersers do not today overlap with the ranges of close to 500 of these species. Madagascar is home to increasing numbers of orphaned plants.

Habitat destruction is the single biggest threat to Madagascar's wildlife, but it is not the only one or perhaps even the most important for some species. Hunting to put food on the table or supply the bushmeat trade is on the rise in the face of hunger, the breakdown of traditional authority in the countryside, and demand from city restaurants. *Fady* eventually came to protect many of Madagascar's animals. An animal that is *fady* cannot be eaten, or even touched in the case of tortoises in the south, and any violation risks incurring the wrath of ancestors or other spirits. But not all animals are protected this way. Besides, *fady* are usually specific to a particular region and ethnic group. Newcomers may not respect prohibitions, and urbanites on hunting expeditions in the countryside may ignore or be unaware of them.

Reliable evidence about hunting and bushmeat consumption is hard to collect, particularly for species hunted illegally. Fearing punishment, people are unlikely to give honest answers to questions. Using methods designed to get around this, a recent study inquired of villagers in six communes across the island whether they had eaten the meat of lemurs, *fosa* (*Cryptoprocta ferox*), wild ducks, fruit bats or tenrecs in the past year. All except one of the communes was within a few kilometres of forest, and two bordered long-established, formally protected areas.

Tenrecs are classified as game that can be hunted at certain times of year. According to the study, they were by far the most commonly and widely eaten, with 70–90 per cent of people reporting that they had done so. The two other game species, wild ducks and fruit bats, were also eaten quite commonly. About a quarter of people reported eating lemurs in communes with no protected area nearby, and a much lower proportion in communes bordering protected areas.

These findings do not reveal how much or how often people ate bushmeat over the previous year, and the sample size is small. But they are certainly cause for alarm, particularly for tenrecs. Even with their stupendous reproductive powers, it seems unlikely tenrecs can withstand high and sustained levels of harvesting.

Demand for some species comes more from outside Madagascar than within. The ploughshare tortoise has almost completely disappeared from its small geographical range in the west in the last few years, swept up by the pet trade in East Asia. A few foreign professional dealers drive an international trade in less glamorous reptiles and amphibians as well, with tens of thousands exported each year to the United States,

Europe and East Asia. At least 56 reptile and 16 amphibian species are targeted for export, including frogs, small tortoises and turtles, snakes, geckos and chameleons. Valued for their looks or sheer rarity, the small body size of most of these species makes them easy to smuggle, corruption eases the way, and the dealers are canny and hard to stop.

The impact of human activities described so far is direct – habitat destroyed, animals killed or captured alive – but it is also indirect. From the time of early settlers, people have brought new species to the island, including 50–60 animal species and well over 1,000 plants. Some introductions were deliberate while others happened by accident, and the impact might be slow and subtle. But that does not make these introductions unimportant. Consider a few examples: black rats are vectors for disease, and probably compete with native rodents; imported frogs likely brought with them the chytrid fungus that kills its amphibian hosts and is causing worldwide extinctions; a new species of aggressively invasive toads is setting off alarm bells; myna birds displace native bird species along forest edges and in open areas; freshwater fish are declining in the face of new competitors and predators; wild cats in the south and west have joined domesticated dogs as non-native predators on lemurs and other animals. The list goes on.

The varied consequences of introductions need more attention, but meanwhile there are crises playing out in full view. In the spring of 2018, 10,000 radiated tortoises were discovered in a home in Toliara. They covered the floors, jammed up against one another in every room. Like ploughshare tortoises, they were destined for pet markets in East Asia. Big shipments have been intercepted in Bangkok, and the trade has decimated the population of radiated tortoise once abundant in the south. I have experienced this firsthand.

*Captured radiated tortoises (*Astrochelys radiata*) in Toliara*
(Turtle Survival / Jordan Gray)

It is a long step up into a Land Rover. The chassis sits high over the wheels, which is helpful when rolling along an unpaved road strewn with rocks. You just drove over them, and they rarely forced a halt on the unpaved National Route around the southern rim of the island. Tortoises were the main problem, rather – until the last decade. I cannot tell if very big radiated tortoises decided to trundle across the road when they heard the sound of an approaching car, though it sometimes seemed as if they did. The sight of a large obstacle ahead would regularly bring us to a lurching standstill. One of us would get out, pick up the tortoise urinating its displeasure, deposit it in the adjacent bush, and on we would go. Those days are gone. There are no tortoises on the road today.

*

This chapter began with a patch of forest that disappeared, replaced by maize fields. People undertake the back-breaking work of clearing forest for a variety of reasons. Widespread clearance in the southwest for maize cultivation over the last few decades illustrates this and makes a broader point: although rural farmers wielded the axes, behind them were ranged a policy initiative in Europe, Malagasy and overseas governments scrambling to take advantage of it, pressure from international aid agencies, and investors and businesses seeing an opportunity for profit. Each axe was powered by many hands, many interests. Poverty surely played a role, but not necessarily the only or defining one. Indeed, having more money makes it possible to buy better tools and pay more people to clear more land.

Maize cobs were a common sight during harvest season when I began working in the southwest, spread out on mats, dangling from eaves or festooning the roofs of huts as they dried in the sunshine. Cobs were part of the village landscape, grown mostly to feed the family, with any surplus traded in local markets. They always had a cheerful air, I thought, betokening a good harvest and food for the coming year. Then came a maize boom and with it the rapid conversion of forests to fields. The cobs became a mountain. Immigration from other parts of the south, the difficulty of growing most other crops and lack of alternative livelihoods all contributed to the boom, but changes in international politics and economics were its main driver.

In 1990, the European Union established new policies to stimulate economic development in former colonies. The island of Réunion, an overseas department of France, lies about 1,000 km east of Madagascar. Taking advantage of the new policies, the government of Réunion targeted agriculture and offered tax breaks on imported grains in order to make animal feed cheaper and support meat production. Madagascar was nearby, transport costs were low, and orders could be met quickly. Springing into action, the Malagasy government built a storage silo on the coast and set up a maize collection system. With dramatically improved access to a hungry overseas market, prices quadrupled. Maize was now a lucrative export commodity, serenaded by axes clearing land for more. The boom ended in 2000 as abruptly as it began and, again, the causes lay outside Madagascar. Réunion built a port and storage facilities that could handle huge, cheaper shipments produced by international agri-businesses, and was soon importing almost all its maize from them. The legacy of the boom endures in Madagascar, however. The network and infrastructure it built shifted to national markets. Now, maize from the southwest is turned into chicken-feed for the capital's poultry industry.

These events are a saga of people in the countryside working hard to improve their lot; a well-intentioned EU policy initiative; an enterprising response by the Malagasy government of the time; and the seizing of an opportunity to increase profits by businesses and investors. Most regions in Madagascar have at least one experience of this kind, and often more. By 2009, there were foreign investments in over 3 million hectares of arable lands and large-scale international mining investments as well. Although the details differ and do not always involve international interests – demand for charcoal in southwestern towns and cities drives another whole supply chain that begins, again, with small-headed axes – the saga invariably entails forest clearance and dwindling wildlife populations. Whether legal or illegal, most exploitation begins with one person wielding an axe. But that is not the whole story.

Over the last 50 years, a new group has joined the fray with an altogether different interest in axes. Under colonial rule, environmental protection took second place to exploitation and attracted little support. The plight of forests and wildlife began moving to the forefront with the first major international conference in 1970, however, and conservation efforts really took off in the mid-1980s. Public perceptions were changing and western environmentalism was on the rise, fuelled by powerful works like Rachel Carson's *Silent Spring*, published in 1962. Centuries of real and imagined perceptions of nature in Madagascar on the part of European visitors made it fertile ground. A British environmentalist coined the now widely-used phrase 'biodiversity hotspot', and identified Madagascar as one of a handful of global hotspots. This new prominence attracted funds and interests quite distinct from those of human survival or commerce. Conservationists celebrated.

Financial support came overwhelmingly from international aid agencies, the World Bank, non-governmental organisations and foundations in Europe and the US, and so did many of the people effectively deciding how the money would be spent. Between 1990 and 2009, the National Environmental Action Plan mobilised almost half a billion dollars from outside the country. It was a staggering amount. Some have characterised this largesse as little more than neo-colonialism or a 'green grab' by foreigners. The reality is more complicated, and dichotomies stressing the ascendancy of foreign over local interests ignore the part played by institutions and individuals in Madagascar itself. The Malagasy government's promotion of environmental policies was deliberate and far-sighted, not simply an opportunistic exercise in extracting money from rich foreigners, and a growing cadre of well-trained Malagasy conservationists lead many efforts today. Environmental advocates both within Madagascar and beyond joined forces to become serious participants in the struggle for control of the axe.

*

Governments of Madagascar have been making efforts to protect the island's forests for over two hundred years. They have not met with great success. Protection is hard to ensure and, at least in the short-term, clashes with the goals of meeting basic needs for food and fuel, and of increasing wealth and prosperity. This was evident in the actions of King Andrianampoinimerina (see also Chapter 8), hailed by some as Madagascar's first conservation-minded leader. On the one hand, the king did indeed prohibit exploitation of woodland on the twelve sacred hills of Imerina; on the other, he encouraged forest clearance to create the great rice irrigation system that would provide food for his subjects.

In the early twentieth century, the colonial government established natural reserves to protect a few tracts of forest around the island even as it encouraged forest clearance everywhere else. Rural Malagasy cultivators were by now identified as villains, with their 'barbarous and deplorable practice' of 'slash-and-burn agriculture', and several laws were passed to stop it. In practice the ban did not work, and the illegal burning of forests and grasslands also became an enduring symbol of protest against the state.

Conservation gradually emerged as a major priority over the last 40 years through a stream of policy developments that took place despite frequent political turmoil. On paper, there was great progress. Madagascar was one of the first countries in the world to develop a National Environmental Action Plan, with successive phases spanning 20 years. The Plan had far-reaching goals in education, health and sustainable development as well as environmental protection. A National Park Service quasi-independent of government was established, quickly followed by a Ministry of the Environment. In 2003 then-President Ravalomanana set an ambitious new goal to triple the country's network of protected areas. It actually expanded fourfold over

1970	International Conference on the Conservation of Nature and its Resources held in Antananarivo, co-sponsored by the Malagasy Government and international institutions
1984	National Strategy for Conservation and Development announced
1985	International Conference on Conservation and Development held in Antananarivo, sponsored by the Malagasy Government
1990	National Environmental Action Plan signed into law and launched with a 15-year timeframe
1991	National Office of the Environment and precursor (known as ANGAP) of Madagascar National Parks established
1994	Ministry of the Environment established; passage of first environmental impact assessment legislation
1996	Legislation facilitating transfer of resource management to community associations; Tany Meva established as Madagascar's first national environmental foundation
1997	Comprehensive new forestry legislation
2001	Additional legislation facilitating transfer of resource management to community associations
2003	Legislation outlining different categories of protection for protected areas; 'Durban Vision', a commitment to triple terrestrial protected areas in 5 years, announced by President
2005	Policy for Protected Area System (known as SAPM) established, with responsibility for individual reserves delegated by the Ministry of the Environment to Madagascar National Parks, regional authorities, NGOs and private actors; National Biodiversity Foundation established with goal of providing sustainable financial support for protected areas

Major national actions concerning the environment since 1970 the ensuing decade to include 7.1 million hectares at 122 sites. Designed with considerable scientific input and expertise, the network today encompasses and in theory protects much of the island's biodiversity.

Policies increasingly emphasised decentralisation, with the delegation of land stewardship to local communities under a series of new laws. Most of the new protected areas were established with complicated governance arrangements shared between local communities and foreign conservation organisations, and authority was transferred entirely to several hundred communities. The shift, echoing pre-colonial ways, made good sense in principle. Rural Malagasy had the most to lose when denied access to forests on which they depended for their livelihoods, and it was fair and right to engage them directly in ways to manage lands of which they were the customary proprietors. Besides, in practice the centralised imposition of protection did not work in a state strapped for money and the personnel needed to enforce it.

The successes of these policies have turned out to be disappointingly few. It is ironic that environmental and conservation interests promoted by people and institutions in many ways so rich, powerful and privileged, have not fared better in the contest for land. None of the goals in education, health and sustainable development have been met. Forest cover continues to decline island-wide. Deforestation has slowed in the network of protected areas as a whole, but the effects are uneven and modest at best. Deforestation rates in forests wholly managed by communities are no different from those managed by the state.

Reasons for these disappointments are many. At the national level, Madagascar has experienced many political crises over the last 50 years, its economy is weak, and the country is one of the poorest in the world according to World Bank rankings. Laws are enforced inconsistently at best, and corruption is widespread. Revenues from international tourism have become a mainstay of support for protected areas, but the catastrophic collapse of these revenues in the wake of the COVID-19 pandemic underlines their unreliability. At the local level, engaging communities meaningfully in resource management has turned out to be easier to advocate than achieve. Community rule regimes are products of arduous local negotiation and compromise, processes dependent on community history and composition. Even with seemingly effective rules in place, community members are often reluctant to apply them against one another and powerless to do so against outsiders. In some places, local elites have captured and controlled management for their own benefit.

Problems with efforts to decentralise conservation efforts have deeper roots still, however. Land has profoundly different meanings and value for people who live and make their living in the countryside. One jaundiced view is that management transfers 'are aimed at involving local communities in achieving natural resource management goals (conservation and "sustainable development") that are not their own, while at the same time pretending that if local communities are given the rights to manage natural resources, these same goals will be more easily achieved'. The impact of climate change brings new challenges. It is easy to despair.

*

What is happening in Madagascar is not unique. A host of environmental losses and a tsunami of extinctions are sweeping the world. By no means the first cataclysm our planet has experienced, this one is different because *Homo sapiens* is its primary cause. Some liken our species to a weed. Plants designated as weeds usually grow rapidly, reproduce prolifically, tolerate a wide range of habitats, and are able to move long distances. They are good at spreading fast and invading new territory. Some animals have weedy qualities too, with additional attributes like curiosity, an aggressive and gregarious temperament, and speed and agility on the ground. The human animal is a weedy species of an unconventional kind. In addition to disrupting the places it invades, it has achieved the dubious distinction of combining prolific reproduction with a long lifespan and, in the western world, profligate consumption. From the standpoint of other plants and animals, we are weedy world champions and the grounds for despair are palpable.

The future prospects of our planet seem a bit less gloomy if one listens to farmers in Madagascar, who have a different view of weeds. They are quick to recognise new, sometimes rapidly spreading plants in their fields and surroundings, unconcerned about their origin, and slow to label them. They withhold judgment, waiting to see if the new plant will turn out to be useful. They call it a weed when it competes with crops and reduces the harvest, but not when it grows in a resting or fallow field. Their view of weeds offers a more hopeful metaphor for our own species. Weeds *can* be flowers.

Writing this chapter was hard because it is largely about loss, unresolved conflicts and failed or faltering efforts to resolve them. The losses are real and it certainly does not help to underestimate the magnitude of the challenges facing Madagascar and yet, taking my cue from Malagasy farmers, perhaps loss, conflict and failure are not inevitable. I found myself wanting to interject evidence of this as I wrote. The successes are small and fragile, but they too are real. What works, and what can be learned from the successes?

Monitoring team member Elahavelo Efitroarana, with a newly painted label for the botanical garden, 2016 (photograph by author)

CHAPTER 11

Places That Work

In 1982, a Canadian film producer asked me why I was wasting my time on conservation in Madagascar since the last tree was already burning. That's not true, I replied. My answer still stands.

Outside the window, rain bucketed down and mist shrouded the forest-covered hillsides. Even inside the meeting room it was raw and chilly, and everyone sat bundled up in sweaters and parkas. Rainforest is not supposed to be this cold, I thought. We were assembled for the opening session of the International Primatological Society's annual meeting at Ranomafana National Park in August 2013. In the audience was a scattering of foreigners like me and several senior Malagasy researchers, most of them colleagues and old friends. But many of those present were young. A few I knew and more I did not. Just seeing a rising generation of Malagasy field biologists and conservationists in such large numbers gave me hope. There were nowhere near enough to fill a big room when I began working in Madagascar. Conversations during the days that followed, hearing about their research and ambitions, sensing their palpable commitment, helped banish despair to the sidelines.

Patricia Wright came to the podium to welcome us. 'I promise it wasn't like this in August last year, the weather was beautiful then . . .' she began wryly, with a broad smile. I silently clocked up another data point for the unpredictability of rainfall patterns in the eastern rainforest! Distinguished primate researcher, Pat is also the indomitable force whose vision and energy drove the creation of Ranomafana National Park in 1991. The Park protects over 41,000 hectares of rainforest and a remarkable array of wildlife, with the further goal of reducing the need of people living around the periphery to exploit its plants and animals. The Park is an international centre for training and research, and nurtured many of the Malagasy students and researchers in the room. Pat had reason to feel proud as well as welcoming.

No, the last tree is not burning. There are still a few great tracts of forest like Ranomafana across the landscapes of Madagascar. Almost all are now designated as national parks or nature reserves, managed variously by Madagascar National Parks and non-governmental organisations (NGOs) – mainly international conservation organisations, although with the increasing involvement of national and local NGOs. The most accessible of these protected areas have become magnets for tourism, complete with visitor centres and well-trained guides.

A hundred and twenty-two protected areas are scattered across the country and surrounding waters today, ranging on land from grand spaces like Ranomafana to small pockets of forest. They constitute 'a dizzying range of forest types and forms of use . . . in a broad-ranging experiment with forest governance'. There have been no attempts to do away with protected areas during the political upheavals of recent decades, and successive governments have in fact more than tripled their number and elaborated the laws regulating them. Overall, the network is effective in reducing deforestation. These are not trivial achievements, even though sometimes more successful in theory than practice.

Amid the anguish about ongoing destruction, it can get lost that Madagascar has an abundance of people who, for a wide variety of reasons, value the natural world around them. It helps provide the necessities of daily life for some and supports a cash income for others. For many, the forest is a special place because it embodies the stories of their lives and is where their ancestors are buried and their descendants will live. There are also growing numbers of Malagasy, like those in the chilly meeting room that day, who share the values of the international conservation community. And they all have an additional reason for caring: Madagascar is their home.

The presence of people does not inexorably doom their surroundings to destruction and nor does cherishing those surroundings inexorably doom people to lives of abject misery. Human livelihoods and local conservation efforts thrive together in some places in Madagascar, under quite different circumstances. How does that happen?

*

Marine environments are easier to protect than those on land, or at least less difficult, because protection more often goes hand in hand with immediate economic prospects and this gives people a tangible incentive for action. Rapid and sustained increases in managed marine populations benefit not only the critters themselves but also those who make a living catching them. A clear demonstration of this in Madagascar took place at Velondriake, where nearby mangroves and coral reefs teem with fish, shellfish and octopus. It is somehow fitting that this is also the site of the oldest fishing village discovered thus far.

In 2004, working with Blue Ventures, an international NGO, the women and men of Velondriake who live by fishing agreed to an experiment: for just a month, they would stop catching the octopus on which the local economy depended. Even during so short a time, the octopus population shot up and so did catches when fishing resumed. This led the fishers to form a management association to govern no-fishing periods. Word of their success spread fast, and 23 neighbouring villages followed suit. Today, there is a string of locally managed marine areas up and down the coast and partnerships with seafood export companies provide a guaranteed market for the catch.

The economic incentive for restraint was quickly obvious to the fishers of Velondriake. It is usually much less apparent on land. Many years go by before a forest grows back after all the trees are cut down. Heavily hunted animal populations do not rebound after a month-long no-hunting period. They stay scarce. The land is less bountiful than the sea. The literal translation of *ala tahiry*, the Malagasy words used to refer to protected forest, is *forest that is set aside*. A forest set aside has little or no immediate economic value. It represents an opportunity cost for people nearby, rather, and their lives do not overflow with alternative opportunities. Therein lies a major dilemma and challenge for terrestrial conservation.

The Tsimembo forests in the west are an exception to this, because of the freshwater in their vicinity. They surround the Manambolomaty Lakes – let us call them the western lakes – which contain an abundance of fish. Local fishing people catch them, and so do endemic Madagascar fish eagles (*Haliaeetus vociferoides*). Historically, a *tompondrano* (keeper of the lakes) decided when the fishing season started and ended and what was allowed. This system of customary rules worked well for fish, people, forests and fish eagles alike. Fish stocks were healthy, the catch was good, the surrounding forest remained largely intact, and fish eagles thrived.

The system broke down in the 1990s with a seasonal influx of fishers from villages as far as 50 kilometres away. Within a few years, the number of people fishing on the lakes increased from about 30 to 300 and dozens of temporary camps sprang up in the surrounding forest. Rules were ignored and the lakes over-fished. Seasonal immigrants used wood from the forest to make dugout canoes and provide fuel for fish-drying fires, disturbing and degrading an important breeding site for fish eagles. It was hard to see how the downward environmental spiral could be reversed. Fishing on the lakes was profitable, and worth enduring the hard work and camping away from home for several months each year in the case of immigrants.

Fish eagle (Haliaeetus vociferoides) *in the western lake complex*
(The Peregrine Fund / Lily-Arison René de Roland)

At this point, The Peregrine Fund (TPF) came on the scene. The primary mission of this international NGO is to save birds of prey from extinction. The Madagascar Director, Lily-Arison René de Roland, has led TPF's activities and research for many years now. He is a big, softly spoken man, whose face lights up with joy at the sight of a long, curved beak and huge, powerful talons – as I discovered when we travelled to the lakes together in 2014. The trip turned out to be overshadowed by *dahalo* (bandits) active in the area at the time. Whole villages had emptied as people fled to camps safely hidden in the forest, and we spent much of our visit talking with local leaders about what could be done to improve the situation. But there was still time to go out on the lake and watch fishers gracefully and rhythmically cast their nets from dugout canoes, each sporting a painted permit number, and to stare up at fish eagles staring down at us with cold eyes from the treetops.

TPF's starting point was that if there were enough fish in the lake and wood in the forest to meet the ongoing needs of people, then there would be enough food and nesting sites for fish eagles as well. Making this a reality called for renewed incentives and firmer controls over the harvest of fish and wood, and TPF set out to help the community strengthen and enforce the failing customary rules. Community leaders (*tompondrano*, mayors and elders) held public meetings to announce a *dina*, or local social contract, embodying the rules. A *dina* properly enacted carries great weight, but it was not enough to make a real difference in this instance and two community associations were subsequently established to help organise and catalyse efforts more effectively. The associations, acting on behalf of the community, petitioned the government for the formal transfer to them of management authority, and a boost to the petition came with recognition of the region as a 'wetland of international importance' under the Ramsar Convention. After lengthy negotiations and many bureaucratic delays, the government approved the petition for a three-year probationary period in 2001.

The contract has been renewed several times since then. The community associations have successfully limited the number of local and migrant fishers on the lakes, the catch, the length of the fishing season, and net mesh-size; they have introduced more efficient fish-drying methods that reduce fuel-wood consumption, controlled the harvesting of trees for canoes and construction, and overseen the planting of thousands of tree seedlings to restore forest along the lakeshore. In 2015, a national decree established the western lake complex as a multiple-use protected area encompassing 200,000 hectares of biologically rich forest, wetlands and grassland. The human population remains small and still relies heavily on lake fish for a living, and three communities continue to manage the fish and forests through locally established associations. The number of eaglets fledged rose significantly over the first decade of community efforts, and the fish eagles of Tsimembo-Manambolomaty – about 10 per cent of the entire Madagascar fish eagle population – continue to hold their own today. And the fishing families do too.

*

Analafaly is the village in the southwest I described walking or wading to in Chapter 2, in one of my many encounters with unpredictable weather in Madagascar. For me, unpredictability makes the difference between easy and unexpectedly hazardous excursions to Analafaly. For the people who live there, it makes the difference between having enough to eat and going hungry. The village lies a few kilometres east of the Bezà Mahafaly

Special Reserve. The reserve encompasses riverine and spiny forest, and is home to an abundance of wildlife including the *sifaka* I have spent many hours watching.

It was lunchtime in January 2014 and Joelisoa Ratsirarson was standing in the blazing sun outside the Analafaly primary school, with a bemused smile. We already encountered Joelisoa among the plant experts wandering across grasslands in the southwest. On this occasion, however, he was surrounded by a sea of children each clutching a battered tin plate and mug. I hung out gratefully in the shade of an acacia grove with friends from villages nearby, chatting idly as we surveyed the scene. Beside us, women squatted to tend open fires over which bubbled vast cooking pots of rice and beans. The milling children were cheerful, but they were also hungry. Last year's harvest had all been eaten, sold or used as seed, and this year's crops were not yet ripe. It is always a hungry period, and famine looms if the previous year's harvest is poor. Like the last few years, this was such a year. With little energy, many children had abandoned the daily walk to the schoolhouse and attendance had plummeted. Community leaders asked for help. That was the beginning of the school canteen in Analafaly and other villages. It takes about 10 kilos of rice and 5 kilos of beans to provide a meal for 80 children. Joelisoa had found funds to buy rice and beans until the new harvest was in, and mothers took turns to cook for the children each day. We were there to ask how things were going.

This is conservation? It is. A professor in the School of Agronomy at the University of Antananarivo, Joelisoa Ratsirarson has led research and training at the reserve for over 20 years. He is no ordinary academic. Alongside his career commitments, he and his wife, Vololona Rakotozafy, have poured their energies into making conservation work at Bezà Mahafaly.

The canteens were one more small action in the long history of a partnership forged between community leaders and a handful of academics from Madagascar and overseas, including myself.

Immediate economic incentives for conserving the forest at Bezà Mahafaly were not at all obvious in 1975 and nor are they very obvious today. People do have reasons for valuing the forest, to be sure. The presence of ancestral tombs makes it sacred, and in fact the name of Analafaly village means *place of the sacred forest.* The forest offers pasture for livestock or concealment from cattle-rustlers, and is a source of honey, fruit, medicines, thirst-quenching tubers, trees to make planks, and dead wood with which to cook and keep warm during chilly nights. At the same time, it is a potential site for fields, and the food security and income they offer. The scales are finely balanced between competing needs.

The partnership started out as a bargain struck: in exchange for the community's help in keeping the forest safe, we academics would seek external funding to help the community achieve goals of its own. Each side had its own interests and priorities. The community stuck pretty faithfully to their side of the bargain, designating small patches within the larger forest for total protection to create the reserve in 1986. They supported its subsequent expansion, and established a local conservation *dina* to help enforce rules of access and use. We academics, on the other hand, struggled to honour our commitment. Gradually, though, benefits came. Wells were dug, primary schools built, adult literacy classes organised, seed stocks improved. Small incremental changes, they brought no dramatic transformations to village life but made a difference for the better in the eyes of the community.

As the years went by, honouring a bargain came to feel more like being good neighbours. Perceptions changed. 'We' were

241

surely as mystifying to 'them' as they were to us at the outset, but those categories slowly broke down. From the amorphous notion of 'local community' emerged individual men and women, old and young, well-to-do and poor, with their own alliances and conflicts; academics returning year after year became individuals appreciated for their commitment, known for their particular interests and – in the case of foreigners – admired or quietly laughed at for their skills (or not) in the Malagasy language. Trust built. In 2004, Madagascar National Parks joined the partnership, elaborate mechanisms of co-governance and management were established and, with bumps along the way, a growing, shared sense of purpose superseded the original 'exchange of goods and services' arrangement.

There were worries, though. Tangible benefits to the community were entirely dependent on external grants and would disappear if funding dried up. What could be done about this? Jeannin Ranaivonasy inadvertently provided one possible answer when he loaded a big sack into the car one day in 2011. Like Joelisoa, we caught a glimpse of Jeannin on the southwestern grassland excursion. Jeannin was deeply involved with the Bezà partnership for many years, and used satellite images to track changes in forest cover in the reserve and surrounding region. His sack had nothing to do with this, however. It was full of locally produced *siratany* – meaning salt of the earth. *Siratany* contains naturally high levels of potassium chloride and relatively low levels of sodium chloride. It is deemed a 'healthy salt' in Madagascar, and the sack was on its way to Jeannin's grandmother who suffered from high blood pressure.

Siratany comes from patches of salty soil formed in ancient times when the region was under the sea. Women scrape soil from the surface of these patches and mix it with water to produce a soupy brine, which they strain through a tightly-woven

*Siratany cooking near Analafaly, with earth from which brine was extracted
in the background, 2017 (photograph by author)*

reed mat suspended over a hollowed-out tree trunk. The filtered
brine is cooked in open pans for several hours, and then the
siratany is ready for final drying. The production process is long
and arduous, yet the going price in the local market is just a
few pence per kilo. Today, working with the team, producers
have developed a new form of hearth to cook their salt;
constructed from rocks and mud, it uses far less wood than the
old hearths. The village community has also set aside open land
and planted tree saplings there, with the goal of providing wood
fuel over the long-term. Hopefully, helping link the producers
to new, broader markets will prove an enduring way to strengthen
a sustainable and largely ignored thread in the local economy.

The friends standing in the shade with me that scorching day

243

at Analafaly have been catalysts for this and many other activities. They are members of the longstanding, locally-recruited Bezà Mahafaly monitoring team, who collect systematic data year in, year out that is gradually building into an important environmental record for the area. Today, this information is stored in a database jointly stewarded by the School of Agronomy in Madagascar and the Yale Peabody Museum. Over the years, team members have also become trusted and knowledgeable intermediaries, explaining community concerns and aspirations to the academics and Madagascar National Parks staff and vice versa. The team helps coax collaboration along.

What does success look like at Bezà Mahafaly? Jeannin's analysis of satellite images provides one demonstration of the partnership's effectiveness. While unprotected forests have declined significantly over the last 40 years, protected forests in and immediately around the reserve remain largely intact. Communities neighbouring Bezà Mahafaly have come to appreciate the value of the Bezà conservation *dina*, and seek advice on how to establish one themselves. The conservation *dina* has also become the model for a security *dina* across a broad swathe of the southwest, following a breakdown in law and order after the military coup in 2009. At meetings these days, updates on the *dahalo* situation mingle with talk of ways to safeguard forests and wildlife, and plans are afoot to expand the security *dina* to include conservation. It is an exercise in scaling-out rather than -up.

*

Sarah Osterhoudt used the lovely phrase 'happy landscapes' to describe the forests around the northeastern village of Imorona. Her account of the reaction of Imorona farmers to learning that vanilla is not native to Madagascar featured in an earlier chapter, and I return to the village now. In the places we have visited so far, communities and external organisations came together and developed a shared goal: protect the environment while enhancing local livelihoods. The farmers of Imorona achieved this by a different route, under their own steam. The village sits in marshy lowlands where farmers grow rice, and a patchwork of mature forest, secondary forest and agroforestry fields covers the hillsides behind the village. Trees cover more than half the community's land. Agroforestry fields are a far cry from the monotonous spaces blanketed with a single crop that come to mind when I think of fields. They are complex systems, mixes of cultivated and wild trees and bushes, annual crops, livestock and even small fisheries. Fruit trees – lychees, mangos, jackfruit, oranges – stand alongside trees grown for firewood and construction, herbaceous plants for food and medicine, and coffee bushes, clove trees and, most precious of all, vanilla vines grown as cash crops. Sarah recorded 97 tree species growing in fields around the village, over half native to Madagascar, and her records also included the Malagasy names of 73 types of herbaceous plants, representing at least 67 scientifically recognised species.

The agroforestry practices of Imorona are decades or centuries old rather than the outcome of recent collaboration with an NGO, but they did not develop in isolation. People settled the region many centuries ago, and it has a long history of overseas contacts and trade. Portuguese, Dutch, French and Chinese merchants all did business there, and the regional port of Mananara even served briefly as a kind of promised land for pirates – indeed, many families today count pirates among their ancestors. These external contacts brought new crops and ideas, and relative prosperity. Most farmers maintained their fields as diverse, managed forest landscapes and incorporated new crops and ideas they found useful. They accepted vanilla and coffee but not sugar and peppercorns, declined substantial cash offers for land, and ignored the urging of French colonial officials to develop plantations containing only a single crop. In recent decades, volatile commodity markets, the devastations of frequent cyclones and allure of growing urban centres have increased the pressure to walk away from historic ways of working the land. A 30-fold increase in vanilla prices in the last few years has brought new wealth but also new pressures and tensions. On the one hand farmers have more to gain but on the other they have more to lose, and theft has become common, with suspicion, fear and even mob violence travelling in its wake. Yet by and large the system endures through these vicissitudes. Why?

Vanilla pods (Vanilla planifolia) *etched with their owner's name, 2018 (photograph by author)*

(*Above*) Black and white lemur, *Indri indri* or babakoto, singing; (*top right*) Crested coua, *Coua cristata* or tivoka, in all its glory; (*right*) Brown leaf chameleon, *Brookesia superciliaris* or ramilaheloka, on the move (three photographs by Ken Behrens); (*below*) Baobab, *Adansonia sp.* or za, standing sentinel (Missouri Botanical Garden/ Jeremie Razafitsalama)

Sifaka at Bezà Mahafaly, 2017 (photograph by author)

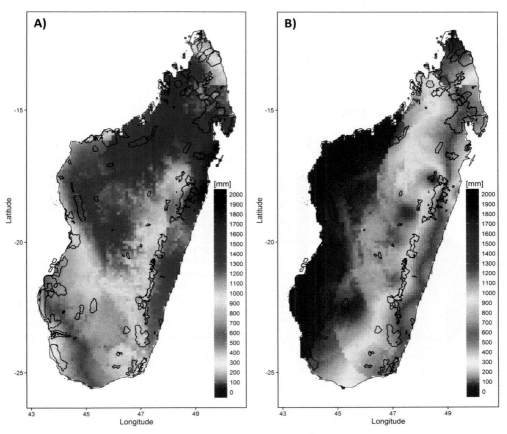

Average seasonal patterns of rainfall, 1985–2014:
A) November to April and B) May to October (Rakotondrafara et al. 2018)

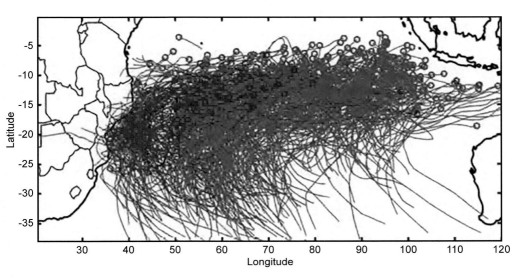

Cyclone tracks in the Indian Ocean, 1952–2007, with those that
made landfall in Madagascar coloured red (Mavumé et al. 2009)

Lavaka in the central highlands
(photograph by Ronadh Cox)

The Betsiboka River disgorging into the Mozambique Channel
(NASA Image Collection/Alamy Stock Photo)

(*Top left*) Tenrec, *Hemicentetes semispinosus* (Afrosoricida) (photograph by Ken Behrens); (*top right*) Fruit bat, *Pteropus rufus* (Chiroptera) (photograph by Harald Schütz); (*right*) Giant jumping rat, *Hypogeomys antimena* (Rodentia) (photograph by Solohery Andrianarivelosoa Rasamison); (*below*) Fossa, *Cryptoprocta ferox* (Carnivora) (photograph by Claudia Fichtel)

(*Opposite top*) Coquerel's sifaka, *Propithecus coquereli* (photograph by author); (*opposite bottom*) Diademed sifaka, *Propithecus diadema* (photograph by Ken Behrens); (*right*) Blue-eyed black lemur, *Eulemur flavifrons* (Duke Lemur Center/ David Haring); (*below*) Ringtailed lemur, *Lemur catta* (photograph by Chloe Chen-Kraus)

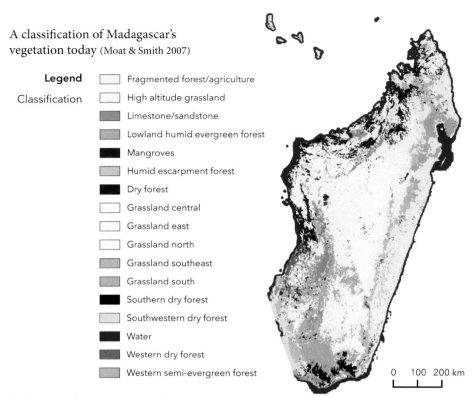

A classification of Madagascar's
vegetation today (Moat & Smith 2007)

Legend

Classification

- Fragmented forest/agriculture
- High altitude grassland
- Limestone/sandstone
- Lowland humid evergreen forest
- Mangroves
- Humid escarpment forest
- Dry forest
- Grassland central
- Grassland east
- Grassland north
- Grassland southeast
- Grassland south
- Southern dry forest
- Southwestern dry forest
- Water
- Western dry forest
- Western semi-evergreen forest

0   100   200 km

Eastern rainforest (photograph by Ken Behrens)

Southern spiny forest (photograph by Ed Lowther)

Palm-studded grasslands on the Horombe plain (Missouri Botanical Garden)

Outrigger canoe under sail off the west coast
(Blue Ventures/Garth Cripps)

Lokanga player at Hazafotsy, with dancers in traditional dress, 1971
(photograph by author)

Colony of elephant birds (*Aepyornis maximus*) on a southern beach; reconstruction based on the density of recovered eggshell and the anatomy of the birds themselves
(artwork by Velizar Simeonovski, in Goodman & Jungers 2014)

Swidden cultivation on steep slopes in the east, 1985 (photograph by author)

Weighing the catch, Manambolomaty Lakes
(photograph by Lily-Arison René de Roland)

Market day in Mahazoarivo, near Bezà Mahafaly, 2018 (photograph by author)

Dancer at inaugural celebration of the Bezà Mahafaly Special Reserve, 1986
(photograph by author)

Festival parade in Imorona
(photograph by Sarah Osterhoudt)

Terraced rice fields in the central highlands
(LouieLea/Shutterstock)

Agroforestry fields certainly offer food security and an income, a very substantial one in boom times, but their persistence is still difficult to explain on economic grounds alone. There are easier ways of profiting from the land. But fields are not simply a matter of economic interest for Imorona farmers and their families. They also embody values, metaphors and meanings, flexibly absorbing cultural as well as material changes. They matter too much to abandon. One of their roles is as a cultural archive of personal and historical events – this orange tree was planted by a grandfather to honour a birth, that clove tree recalls students using crushed clove seeds as ink when ink supplies ran out during World War II. Fields are also 'up-to-the-minute' and the stories embedded in them change – a rock in the middle of a field described by its owner as a place to communicate with the ancestors becomes an illustration of how all entities are under the dominion of God when its owner converts to Christianity. Fields are a place to ponder the future too, for people are aware that memories of them will be associated with their fields and that the condition in which you leave fields when you die is directly linked to the prospects of happiness in the afterlife: 'messy landscapes . . . make for grumpy ancestors'.

Imorona is not a place of pristine forests, and tending agro-forestry fields is not conventionally viewed as a conservation activity. Yet farmers asked by Sarah about the usefulness of trees replied that they provide cash crops and materials for weaving, dyes, brooms, canoes, fences, rope, and ceremonial purposes; and, they went on, forests also offer habitat for birds, food for lemurs, enrich the soil and prevent it from washing away. You could mistake them for state-of-the-art conservationists, except it would not be a mistake. I include Imorona here because it is a place that works on its own terms, and because it illumi-nates the crucial importance of place and history in Malagasy stewardship of the land. *Tontolo'iainana*, the Malagasy word for environment, is a recent addition to the vocabulary, but *tany*, land, is not, and stewardship of *tany* is deeply embedded in Malagasy culture and practice.

*

The small communities at the heart of this chapter have different histories and live in very different environments – a western shore with rich marine life nearby, a mosaic of lakes, wetlands and forests in the interior, an arid landscape in the southwest, and a green, well-watered world in the northeast. The success of community-based, grassroots conservation efforts is difficult and complicated to measure. No matter how measured, it is always fragile.

Some threats to the community activities described here come from outside. Global changes in climate present a long-term challenge everywhere, and particularly in the south. Disparate human interests struggle for control over the land and its resources. General lawlessness brings further perils, from illegal harvesting of timber to cattle theft. Madagascar is poor, and the sustained financial support needed for most grassroots efforts comes primarily from international agencies and foundations. Too often, their expectations of spatial scale and deadlines fit poorly with the particularities of place and time frames on the ground. External funding is increasingly hard to come by, moreover, as attention and resources shift to responding to global climate trends. Ecotourism offers an alternative in principle, but with 225,000–375,000 visitors annually in recent years until COVID-19 drove them to zero, Madagascar's tourist industry is small at best. Many tourists visit with the primary purpose of seeing wildlife – although some come for a beach holiday in the sunshine! – but ecotourism will provide revenue to only a few places for the foreseeable future, and none in times of crisis.

The fragility of success comes also from within the community. A wetland that could be converted into rice fields or a forest into maize crops is a tempting prospect for an ambitious farmer or a family without land; social cohesiveness and resolve are needed to hold on to the wetland or forest, and a change in local leadership or conflict within the community can put these in jeopardy. Conservation *dina* work in some communities and not in others because of complex differences in their histories and in how the *dina* was enacted in the first place. Some contend that the whole idea is no more than the hijacking of a form of social contract deeply embedded in Malagasy culture by government or conservation NGOs for their own purposes. I disagree, though there are surely 'paper *dina*' as well as 'paper parks' in Madagascar.

Against this backdrop, it is reasonable to ask if the 'success stories' are more than cheerful anecdotes of minor interest in a saga of despair. My answer is that they can be, but several things need to happen – and some already are. At the national level, most crucial is the rekindling of honest, competent and stable government, to provide vision and policy, environmental laws and law enforcement, finance and technical assistance. Desperately needed large-scale economic development must be sensitive to its environmental impact, a way out of the fuel crisis must be found, the protected area network needs more investment, and tenure rights to the land must be made clearer.

Changes must happen at the community level too. Community-based endeavours are by their nature small. Today's small, scattered activities must be scaled up or, better, scaled out if they are to have an impact across whole landscapes. Scaling-out, or replication, is most obvious at Velondriake, where a single experiment yielded such positive results for fishing families that other communities adopted their approach, but it is now happening at Bezà Mahafaly too. External partners have encouraged and supported the process of scaling-out in both these contexts. Matters have proceeded the other way around in the case of agroforestry systems like Imorona, which emerged long before ideas about integrating conservation with development came on the scene. With fields demonstrably providing food security and income without destroying the forest, locally nurtured agroforestry systems are becoming models for conservation and development organisations working in eastern Madagascar and across the tropics.

In all the communities described here, people engage in trade in addition to putting food on the family's table. Access to markets is a necessary component of trade, and improving access to markets can be seen as a form of community development. Linked with environmental goals, it becomes a form of conservation. Traditionally, conservation NGOs have had little interest in adding trade and the private sector to their nexus of activities and participants. Concerned about the inequity of helping some families or cooperatives in a community to prosper and not others and about the profit-seeking goals of business, they have viewed the private sector more as a threat than an asset, much as anthropologists have viewed markets as the death of local culture. But this is changing. There are 'good actors' as well as bad in the private sector, the number of good actors is growing and new connections are being forged between markets and communities. Although lemurs are called the 'conservation flagships' of Madagascar, enlisting octopus, fish, *siratany* and vanilla in support of conservation goals may be an important route to ensuring that the flagships themselves survive.

Effective government, scaling-out, long-term external financial inputs, and stronger local economic incentives are all essential ingredients of successful grassroots conservation in Madagascar. Although far easier said than done, they offer a way forward. But, even if 'done', they are not in themselves sufficient. The stories we tell ourselves and one another need to change as well.

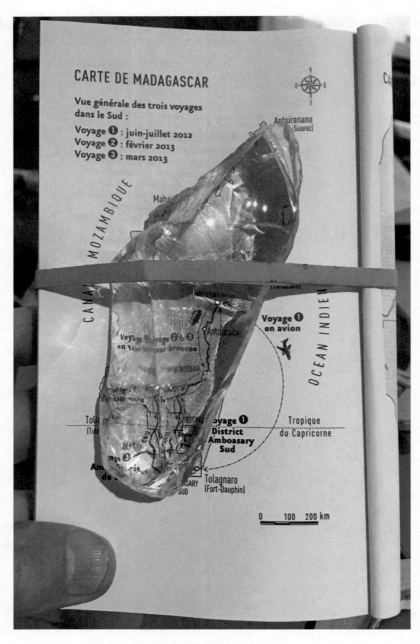

Madagascar through a quartz lens (composition by Ben Gaskell)

CHAPTER 12

Stories Old and New

It was a beautiful sunny morning in June 2018, our boat was bouncing through a gentle swell, and Hervé Andrianjara, newly appointed Director of Masoala National Park, was beaming. We were making our way south along the coast of the Masoala Peninsula, a great finger of roadless land draped in green that juts out from the island's northeast coast. It is in the wettest region of Madagascar, with as much as 7 metres of rain falling in a single year. No wonder things grow well there. Craggy, remote, sodden, the peninsula had the air of a mighty fortress as we followed the shoreline. People have breached its ramparts, however, and in recent decades rosewood harvesters have been hard at work in more accessible areas. Chugging along at the start of our journey, pale green patches signalling fields carved from forest above villages came into sight. As we reached the boundary of the 230,000-hectare National Park, the villages petered out. Now we gazed at a vast, dense forest from sea to mountaintop, and Hervé smiled some more.

The boat headed straight at a sandy beach and slid to a halt in shallow water. I leapt awkwardly off its lurching prow and waded ashore dripping, my mind on those settlers who arrived under sail with no sandwiches and bottled water packed for lunch, no rain gear at the ready, no knowledge of what they would find or how they would live. The line between beach and forest was abrupt. Stepping from sand to leaf litter, we entered a massive green cathedral. No, the last tree is not burning in Madagascar. There are still great tracts of forest like this one.

Our day ended with supper in the beautiful, wood-frame restaurant of an ecolodge far south along the coast of the peninsula. A dog-eared copy of *The Age of Consent* by the British environmentalist George Monbiot lay at the end of the table, and we ate to the sound of waves rolling ashore and rain pouring down outside. That night, the sheets on the bed in my cabin were crisply laundered, and an enormous spider was waiting on my pillow. With a well-timed lunge I smothered it in a towel and returned it to the forest. I do not know what species it was, but based on my description the next morning Hervé assured me it was entirely friendly. The encounter added a certain *frisson* to the ecotourist experience. What a great place, I thought; there really is a future for ecotourism here. And I still think that.

Months later, reading anthropologist Eva Keller's account of the history of the park and the perceptions of communities around it, I learned something new. Reflecting on the ecolodges, a very old woman called Maman' I Jao who lived in a village nearby remarked to Eva that there were many, many *vazaha* (foreigners) in that area when she was young: 'Then they left, and now they are coming back; the *vazaha* don't forget the places where they once enjoyed good times'. In Maman' I Jao's eyes, I was simply a descendent of foreign colonists who invaded the land a hundred years earlier, establishing logging concessions at the site of the ecolodge. It was a wrenching role into which to find myself cast, as if she and I were caught up in two completely different stories.

Stories matter. They reflect and organise the way we view the world, and shape our ideas about the past and the future. Early in the twentieth century, *vazaha* developed a story about Madagascar that tells of a timeless, forested paradise destroyed by human settlers. It still permeates many perceptions of the island's history, including those of some Malagasy. A distinguished French botanist, gifted writer and landholding colonist played a big part in its formulation, and exploring his life and work helps lay bare its roots. If scientific evidence alone were sufficient to overturn it, research in recent decades would already have done so. But this has not happened, and the western world's long fascination with the idea of Paradise on Earth helps explain the story's persistence. A story loses power when a better one replaces it. What might replace this one? My answer is a tapestry of stories, told by many voices.

*

Seeds of the colonists' account were planted in the 1890s. The French government declared a protectorate over the island in 1894, but Queen Ranavalona III and her government refused to buckle and a French expeditionary force was sent to occupy the capital. Rioting followed, and in 1896 France declared Madagascar a French colony and deported the queen, first to Réunion Island and then to Algeria where she lived for the rest of her life.

France placed heavy demands on the new colony: it should be economically self-sufficient and a source of cheap commodities, with Malagasy labour used to achieve these goals. This proved difficult. For one thing, rice production and slaving had been the economic base of the state, but the international slave trade had collapsed and the French government's first act the day after declaring Madagascar a colony was to abolish slavery within the country. For another, the Malagasy did not take well to colonisation. Educated people in the capital fought legislative rearguard actions and nationalist movements took root in secret. In the countryside, widespread resistance and sporadic riots made it hard to enforce the brutal system of forced labour for road and railway construction that replaced slavery. Still, there was already brisk trade in cattle to neighbouring islands, and gold and other minerals looked like lucrative new sources of revenue. Forests did too, and they provided a strong motive for developing an account of Madagascar's history that emphasised the destructive ways of local people and justified the expropriation of forested land.

Some cattle went less willingly than others
(courtesy of Edward Maggs)

A young man called Henri Perrier de la Bâthie shipped out to Madagascar in 1896, the very year France declared it a colony. Working for a French company, his mission was to find the rich deposits of gold rumoured to be present. I find no indication that he did and, in a remarkable feat of transformation, over the next decade the hard-scrabble prospector morphed into an eminent scientist and professor. He also acquired lands at the centre of a big French investment in industrial rice production in the north-west, where he lived much of his adult life. Dubbed 'settler as well as scientist' by the Governor General of the colony in 1924, he became an influential figure in high places in Madagascar and was showered with honours, lionised by younger colleagues back in France, and extolled in an adulatory obituary by Henri Humbert, himself a distinguished botanist of Madagascar.

Contemporary accounts give little sense of what de la Bâthie was like as a person, however. For that, one turns to his monu-mental work, *La Végétation Malgache*. It is full of clues. Meticulous botanical evidence, interesting observations about natural history, frequent references to opportunities for economic exploitation and a strong point of view all leap from its pages in sparkling and evocative prose. It is a life on show, and easy to see why its author meets with a mixed reception today.

With 226 pages of text, tables and photographs, *La Végétation Malgache* offered an authoritative description of the island's flora and is still considered a classic. A wrenching narrative of destruc-tion unites the volume. Perrier de la Bâthie drew a clear bright line between 'virgin' formations and those modified by people, and documented these differences on extensive travels. He cata-logued plant families, genera, and sometimes species, carefully noting the location of each 'virgin' forest visited – though not, tellingly, the sites with open habitats.

The island's grasslands were composed of very few species,

La Végétation Malgache reported. Not only were they species-poor, the grass species that did occur were found all over the world and had been introduced in recent times. This was the botanical evidence for concluding that Madagascar's grasslands were of recent origin. Surely a man who studied the island's plants in such depth must be correct? He for one seemed to think so. In a hefty tome on Madagascar's geography published in 1902, natural scientist and explorer Emile Gautier argued for the antiquity of the central grasslands, on grounds that the highlands had the worst of both worlds, the poor soils of the east and dry climate of the west. They simply could not support unbroken forest. Not so, countered Perrier de la Bâthie magisterially, 'innumerable facts' showed that the arguments in Gautier's 'fine book' were mistaken. Gautier was not alone in his view. Alfred Grandidier, himself a renowned natural historian, remarked around the same time: 'One has to admit that the [central plateau region of Madagascar] must have always been without trees, but not from the hand of man . . .'. But Perrier de la Bâthie's story prevailed, and one wonders how the views of these other great natural historians were so effectively suppressed.

If grasslands were a recent development in Madagascar, a new question followed. What caused the devastation of native forests? Perrier de la Bâthie had a clear answer: people, with axes and fire. Indeed, people introduced fire to the island in his view. And why did they set fires? Sometimes they had reasons: it encouraged new growth for their cattle, made travel easier, or cleared out recently forested areas for fields. But people also burned 'by simple habit and without any reason whatsoever' or, more ominously still, they were driven by 'a strange and contagious mania that drives every Malgache to set fire to dry grasses'. Reading those words a few years ago for the first time, it briefly occurred to me that the speaker at the conservation conference I attended as a young student and described in Chapter 1 must have read them before making his ironic remarks. Almost certainly he had not. The spirit of Perrier de la Bâthie's words lived on in the public sphere then, and still does.

La Végétation Malgache is peppered with observations and advice, sometimes verging on a harangue, about missed economic opportunities. Frustration with the colonial administration sits alongside disdain and dismay bordering on rage at the Malagasy themselves. The volume's appalling last sentence quivers with emotion: 'The conquest of the island would have no reason, no excuse, if we only came here to continue a mindless destruction, without concern for the future, imitating the Malagasy and their childish behaviour.'

The Cochineal Insect Affair tells us that Perrier de la Bâthie's role in colonial matters went far beyond words. Cochineal insects feed heavily on certain varieties of prickly pear (*Opuntia*) in Mexico, their native land. Prickly pear, introduced to Madagascar in the eighteenth century, spread fast across the south and people rapidly came to depend on it. *Raketa gasy*, as it was called, provided food and water in a harsh landscape. It had another property too, which posed a serious problem for the colonial authorities. With long, ferocious spines, *raketa gasy* thickets formed formidable barriers, penetrable only by those who knew their way through the labyrinth of paths traversing them. For the French military it was an operational nightmare, and for the Governor General of the colony a threat: the impenetrable south was becoming 'a place of refuge for malcontents, and, consequently, a centre of rebellion'.

Cochineal insects were brought to Madagascar in 1924, probably imported from a neighbouring island, and within five years they destroyed all the *raketa gasy* in the south. The speed and completeness of the destruction were astonishing, the ensuing starvation and devastation catastrophic. Perrier de la Bâthie published an article about the insects' arrival in the capital. The purpose of his announcement, he wrote, was to 'alert *colons* of the Toliara region [in the southwest] to the introduction so that they might try it in their turn', and he reported that a consignment of insects had already been sent south to Toliara. What he failed to mention was that he himself had sent it. Did he anticipate the disastrous chain of events that would follow? Perhaps not to begin with, although by 1931 he clearly understood the scale of the disaster and tried hard to distance himself from it.

The codicil to this sorry tale is that from about 1930 until independence in 1960, the role of the south became to produce people, cattle and taxes. No more was said about the cultivation of fertile lands, and even the vision of an industry built on cochineal dye that accompanied the insect's introduction proved a mirage: it turned out that Madagascar's variety of cochineal insect produced no such thing. The colonial administration organised the planting of a resistant *Opuntia* strain, explicitly to provide fodder for cattle but perhaps also to feed people. The new variety spread, but the damage and suffering wrought by the 'experiment' could not be undone.

Was Perrier de la Bâthie a botanist and conservationist, or a willing albeit critical collaborator in colonial exploitation? He was both. He gave the world a wealth of botanical information, and played an influential role in the campaign to establish Madagascar's first protected areas. But his advocacy of conservation was itself a form of colonial expropriation, and his beliefs and deep conflicts of interest shaped what he chose to collect and distorted how he interpreted his findings. His work is the source of today's assertion that 90 per cent of the island's forests have been cleared, even though he actually offered three different estimates over the course of his career. The differences were modest – ranging from 70 to 90 per cent – but the vagueness stands out amid the meticulous detail of much of his work, and nowhere did he explain the methods by which he came up with his estimates because they were, in fact, guesses.

Why does the colonists' story persist today, when evidence to counter it has been around and cogently marshalled for decades? It has the usual trappings of powerful stories for one thing, with a simple plot and a clearly identified set of villains (Malagasy people) and heroes (colonists). Myths deeply rooted in the religious beliefs, images and language of western culture provide another part of the answer. Myths are a distinctive kind of story. 'Fictitious' or 'untrue' appears in all three definitions of myth in the *Oxford English Dictionary*, and the term is widely used in the sense of a lie. Anthropologists view matters differently. Myths are the way people 'reach by the shortest possible means a general understanding of the universe . . . with images borrowed from experience [and] put to use'.

The myths of Paradise and the Garden of Eden have pervaded western culture for almost three thousand years. As medieval scholars laboured to identify the location of the Earthly Paradise and debated its relationship to Heavenly Paradise and the Garden of Eden, medieval mapmakers carefully included Earthly Paradise in their images. Their maps continued to reflect Christian scholarship well into the fifteenth century, but during this time sea trade in the western world expanded and there was a great surge in voyages of discovery; the need for accurate maps grew, and by the sixteenth century theologically-inspired locations were appearing less and less frequently. Theological scholars did not give up so easily, and their debate about the location of Earthly Paradise only faded away well into the seventeenth century. But its legacy lived on.

Philibert Commerson, French naturalist and explorer, wrote to a colleague in 1771 after visiting Madagascar: '[Madagascar is] the veritable promised land for naturalists. It is there that Nature seems to have retired as into a special sanctuary . . .'. In the late nineteenth century, German biologist and philosopher Ernst Haeckel had a whole chapter called 'The Site of Paradise' in his treatise *The History of Creation*. His choice of location was Lemuria, the tropical continent posited to have connected Madagascar and Africa before it sank below the surface of the Indian Ocean. Eighteenth- and nineteenth-century European missionaries to Madagascar are perhaps the most obvious flag-bearers for Paradise at that time. They were subject to intermittent persecution by Imerina rulers, however, and were careful to stay away from proselytising in their writings – on the surface, anyway. But, buried within their texts, the Garden of Eden figured as a subtle, common motif along with other coded religious messages to their flock.

Coming to the twentieth century, Perrier de la Bâthie never mentioned Paradise or the Garden of Eden. What was he thinking, though? Perhaps he meant only what he wrote, and the evocative power of his prose simply fired the religious imagination of others after him. 'Abrupt and violent destruction' had turned much of the island into a place 'of ruin, abandonment, useless and impenetrable bush . . .'. At the same time, Madagascar was a 'sort of sanctuary, where all is mysterious and life intense'. Long passages of soaring, celebratory prose punctuate his 1921 treatise, like this description of the seasons in the eastern forest:

The whole forest seems to form but one being, subject to the same laws, the same rhythms. If it's winter, the forest sleeps and its multitudes slumber, rain falls gently, slowly, all seems dead; only the plaintive wails of indri calling back and forth across hillsides disturb the silence descending from the ponderous, leafy vaults above. If it's summer, the forest awakens and life returns to all; storms and periods of radiant brightness follow one another in quick succession across the sky; flower-laden branches quiver with song and the beating of wings, and life is so intense that the woods become impenetrable.

Religious symbolism haunts this nineteenth-century sketch of a palm tree (Ravenala madagascariensis) *being speared to tap water (Feeley-Harnik 2001) (Courtesy of the Royal Ontario Museum, © ROM)*

Such writing was surely an inspiration, at the very least, for recent popular representations of Madagascar that invoke Paradise or the Garden of Eden directly. Take images conjured in the *National Geographic* magazine. An article in 1967 ran under the heading 'Pacific Paradise in the Indian Ocean', and the opening words of one twenty years later were 'A paradise much praised by early naturalists . . .'. An article in 2009 referred to the limestone landscapes of the west as 'a refuge within paradise', and reproduced Perrier de la Bâthie's whole argument with little change. With 3,000 years of western cultural history behind them, Paradise and the Garden of Eden are still in the air, if not always on the page. And they make for a powerful and enduring story.

*

A neat, new 'authorised version' of Madagascar's history does not sit waiting to be told. The island's environments and histories are too varied to encompass in a single, simple account. Changes through space and time may be the one unifying theme, and are certainly at the heart of this book. Our understanding of the past will surely continue to evolve with further research, but here is what it looks like to me – for now.

Drifting across latitudes, first wedged in the middle of Gondwana and then alone, Madagascar was not always an island or where it is today. Upheavals of Earth's crust, volcanoes, erosion, and intrusions by the sea remade the land repeatedly. Soils eroded into rivers running red into the sea, and at times the landscape blazed with fires ignited by lightning or volcanoes. Madagascar was by turns freezing, scorched, well watered and arid before the modern diversity of climates emerged.

Plant species arrived from near and far over the epochs,

replacing one another as conditions changed and evolving their own distinctive Malagasy character. Grasslands spread around the time they became widespread globally, millions of years ago, and boundaries between forest, woodland and grassland shifted in response to regional and global changes in climate. A parade of animals occupied the changing landscapes. Far in the past, they wandered in from adjacent lands in Gondwana. Fossil remains offer glimpses of dinosaurs, and earlier creatures even more different from any alive today. After an asteroid hit Earth 66 million years ago, virtually every animal species living in Madagascar disappeared. It was an island now, reached only by flying, swimming or voyaging on mats of vegetation across the deep, wide channel separating it from the east coast of Africa. Almost without exception, the native animals of Madagascar today are descended from stoic travellers by sea or air.

Little changed for thousands of years after people first set foot on the land around 10,000 years ago, and human clearing of forests eventually began at different times and proceeded at different rates around the island. The first signs of decline that would end in the extinction of the largest-bodied animals came in the seventh century, long after the arrival of people. A range of purposes drove people's activities – to put food on the table by hunting, farming, fishing and herding, to fuel iron-smelting by charcoal production, or to supply the export trade by mining precious minerals. Imperial economic policies in the eighteenth and nineteenth centuries, colonial policies in the twentieth century, and Malagasy government policies, private sector interests and international strategy in our own times have brought ever-faster changes.

Against the dramatic backdrop of recent human impact, it can get lost that Madagascar has always been a land of change.

But did ancient changes ever take place at the *speed* of those brought about by people? The further back into the past one goes, the harder it becomes to discern the timescale of events. We do not know if dinosaurs, fierce frogs, gentle crocodiles and the like disappeared in a day, a decade, a century, or over thousands or millions of years. The best estimates for when the ancestors of modern wildlife arrived span millions of years, yet each actually came ashore on a single day or a single night. Closer to the present but still long before people arrived, sediment cores show dramatic changes in climate and vegetation on the scale of a few centuries or even decades. Rapid change may not be new to Madagascar. What *is* new, of course, is that this time our own species is largely responsible – and the speed at which changes are taking place may indeed be unprecedented.

Madagascar's long history offers release from the clutches of timelessness shattered by the brief moment of our appearance on the scene. A deeper understanding of the past also matters for urgent, practical reasons. Deciding what is worth conserving is one of them. In particular, Madagascar's grasslands have long been ignored and excluded from conservation efforts on grounds that they are a degraded product of human destructiveness. To the best of my knowledge, grasslands are held up as a target habitat for conservation at just one site, leaving a long way still to go.

For millions of years, Madagascar has been like an island with many islands embedded within it, each with its own distinct history, and particularities of place are as important as ever today. Many ecological processes play out over timescales that exceed long-term observations in the present, and evidence of how Madagascar's islands-within-an-island have changed provides clues to how best to manage them going forward. Records of the past offer insight into interactions between plants and animals over long stretches of time and the consequences of adding people to the mix.

Malika Virah-Sawmy belonged in my pantheon of pollen-grain-counting heroes well before she began making direct connections between the past, present and future. Using fossil pollen, analyses of existing vegetation and climate change predictions, she and her colleagues suggested new ways of approaching conservation in the southwest, a region increasingly exposed to high temperatures and unpredictable and concentrated rainfall. Examining the sensitivity of plant species to climate changes over the past 7,000 years, they found that landscape features, relief, soil type and habitat condition all affected how plants responded to the same climate change exposure. From this they developed a landscape framework spanning a gradient from susceptible plant communities that do not stand up well to climate change to resilient communities that do, where human activities would best be concentrated. This framework, they argued, should be a first filter for potential policy and actions that must then pass through cultural, social and economic filters.

When I set out to write this book, it ended here, with the contributions that a better scientific understanding of the past can make to the future. But reading, writing and attending to other voices gradually brought home to me that it is also a story, my story, drawing on the evidence of research to be sure, but also told from a particular perspective. The book acquired a new ending. For one thing, I became increasingly aware that themes embedded in accounts based on scientific evidence and the language used to convey them can have less to do with the evidence itself than with the way we researchers see things.

The vineyards of the Rhône Valley are among the most beautiful landscapes I know. Scattered amid vivid green fields of immaculately pruned vines that stretch to the horizon sit small clusters of grey stone houses, homes of the French farmers who work the land. This is Burgundy wine country, over a thousand years old. People have altered the landscape utterly, and not a leaf or blade of grass that remains is 'natural'. If not natural, does that mean the landscape is 'degraded'? In a strict ecological sense, it does: the diversity of native plants and animals present before people settled the valley has been reduced and many species eliminated, and the natural productivity of the land is diminished. Few would endorse such a label, though. A better description is of countryside transformed or disturbed, with one set of values ascribed to land replaced by another.

Irrigated fields in the central highlands of Madagascar also count among the sights I love. Scattered among the vivid green, immaculately level terraces that march down hillsides and along valley floors sit clusters of red ochre houses, homes of the Malagasy farmers who work the land. This is Malagasy rice country, a few centuries old. Here too, not a leaf or blade of grass in the landscape is 'natural', and here too, calling it 'degraded' seems entirely wrong. Yet 'transformed' is a word rarely used in Madagascar, where a dichotomy between primary forest and degraded habitats typically frames the way landscapes are described. It is one of many oppositions in the dichotomy ledger. On the 'good' side sit ideas and words like originally, primary, natural, forested, no fire, while on the 'bad' sit modern, degraded, cultural, grassy, fire. Such dichotomies are still widely used in ecology, in fact, and Madagascar is far from the only 'victim' of dichotomous thinking.

Each of these oppositions collapses complexity and ignores particularities of history and place. For example, to which side of the ledger should expanding tree cover in the central highlands be assigned? The abundance of trees in the region has been

268

increasing for several decades, particularly eucalyptus, pines, acacias and fruit trees, and some areas being 'reforested' may actually have been grassland at the time of human arrival. People plant trees for fuel, timber and food, and to stake a claim to land. Most are introduced species. Their value for the conservation of native plants and animals is low, though not zero – and they contribute to soil and water conservation as well as human livelihoods. Yet the dichotomy between natural and degraded relegates them to a conceptual no-man's-land. Taken together, the effect of dichotomies is to put the real world at a distance.

'Slash-and-burn' is the phrase used routinely by many people – Malagasy or foreign, researcher, conservationist or conservation sceptic – to refer to a system of forest clearance and conversion to fields used by small-scale farmers in Madagascar. The label conjures aggressive images of destruction about a form of agriculture that in other regions of the world is more often and neutrally called shifting or swidden agriculture. Beyond its negative baggage, the label is deceptive in that it describes only the first steps in a diverse system, reflected in the several different Malagasy words used regionally for agriculture based on the clearing of forests. It also ignores the many hands often wielding those axes in effect.

Glorifying agricultural methods that have produced forest clearance and erosion in some parts of Madagascar is not my point. Rather, in using a pejorative label to condemn a complex form of agriculture out of hand, judge and jury have already spoken in words carrying the full weight of a familiar, generic story. Many Malagasy farmers understand well that their activities sometimes contribute to deforestation and erosion and jeopardise the future of their families' livelihoods. In the absence of other alternatives, however, the only choice open to them – to stop – is no choice at all. Minds in which those activities are designated slash-and-burn are wittingly or unwittingly already part-way closed.

However thoughtfully and well we describe Madagascar in scientific terms, it is only one way of conjuring the land and its history. Accounts that draw on other forms of knowledge, told from other perspectives, are as important to the future as my own, and different stories *can* sit comfortably and fruitfully side by side. From her research and experience living in the north-eastern village of Imorona, Sarah Osterhoudt offers a vivid and illuminating demonstration of this. She came to make a distinction between self-professed authorities on the area's *histoire* and the tellers of *tantara*. This Malagasy word translates literally as story, as opposed to the term *histoire*, the French word for history. *Tantara* recount tales, paint scenes, and emphasise the supernatural. Partly based on accounts handed down through generations and partly on the memories and personal experiences of the teller, *tantara* are not grounded in dates, statistics and the names of political leaders, the facts that structure historians' reports. *Tantara* cluster around themes – particular places, particular crops, family lineages and so on – and their conception of time and space is far removed from the chronological framework of *histoire*.

The setting and style of conveying *histoire* and *tantara* are quite different too. The *historien* recites a chronology associated with the community, consults frequently with notebooks and expects long, respectful silence as he speaks. Seated on chairs in an orderly way, his listeners and he drink chaste glasses of orange soda together at the end of his lengthy recitation, and time is allowed for polite questions. The *historien* is usually an older man with some formal education and knowledge of French, and this form of presenting historical knowledge probably arose as a result of French influences in the nineteenth and twentieth centuries. The *historien* occupies a position of considerable prestige in the community.

The telling of *tantara,* in contrast, is a rambunctious affair. Those gathered are of all ages, they sit on raffia mats on the floor, and people wander in and out. The session begins with an offering of local rum to the ancestors, and then copious quantities of rum get passed around among the living. By the end, several hours later, people are laughing and sharing stories. Talking and having fun are as much a part of the telling of *tantara* as listening and learning. The tellers themselves rarely have much formal schooling. Their craft is learned, rather, through apprenticeship and dreams. They are known for their speech-making skills and ability to mediate between the realms of the living and the ancestors, talents that make them influential in the political and social life of the village – like but also unlike *historiens.*

Imorona is home to a tapestry of stories that sustain a single community. Could such a tapestry help sustain a nation? I believe it could, and that weaving it is essential to the future. A single account, no matter how accurate scientifically, is not enough: 'for landscapes to be truly sustainable they must not only be ecologically diverse and economically viable, but also culturally meaningful'. In Imorona, where the tellers of *tantara* and *histoire* invoke the past in very different ways, the community hears and respects both. It is, perhaps, an illustration of what is possible.

Common ground between differing worldviews, needs, and forms of evidence is not easy to find, and finding it requires good listening as well as telling. There is a world of difference between the conservation community's view of landscapes as places rich in native plants and animals, abstracted from local economic, social and cultural contexts, and the particular place seen by people who make their living there, whose ancestors

271

Conversation with Krisy at Analafaly, 2015 (photograph by author)

are buried there, and whose children will inherit the land. The idea of an all-encompassing common ground is the stuff of utopian fantasy, to be sure. Conflicts over the ownership and use of land will not come to an end in Madagascar or anywhere else. But it is certain that there is more common ground to be found. How much more? The answer depends on the story you choose to tell yourself, those you choose to listen to, how hard you work at listening and, perhaps, your willingness to imbibe both rum and orange soda.

ACKNOWLEDGEMENTS

My thanks go first to Isabella Fiorentino. In addition to being a research sleuth *extraordinaire* and stern but kindly critic of bad prose, Isabella has contributed ideas and insights at every step along the winding path of this book. But that's not all. She has helped me through very dark times, urged me on, and been a true friend. I cannot imagine working on this book and bringing it to a conclusion without her, and my gratitude is without bounds.

The Yale Office of the Provost provided much-needed financial support for my work in Madagascar and the research for this book, and Yale colleagues have been valued sources of ideas and encouragement. Marion Schwartz, in particular, played a central role in all my activities in Madagascar for almost three decades, and I am deeply in her debt.

During sojourns in Madagascar for almost fifty years, many people have shaped my ideas about what matters and how to think about what matters. Some are no longer with us, notably Elie Rajaonarison, Gilbert Ravelojaona, Alison Jolly, Martin

Nicoll and Bob Sussman. I have cherished memories of many conversations with them all. To those still living, especially Joanna Durbin, Frank Hawkins, Pete Lowry, Sheila O'Connor, Simon and Ange Peers, Mark Pigeon, Chantal Radimilahy, Mamy Rajaonarison, Jean Aimé and Vicky Rakotoarisoa, Volololona Rakotozafy, Guy Ramanantsoa, Jeannin Ranaivonasy, Nanie Ratsifandrihamanana, Joelisoa Ratsirarson and Lucienne Wilmé. I offer deep appreciation and gratitude for your wisdom and friendship. My special thanks go also to current or former members of the Bezà Mahafaly team, particularly Efitiria, Elahavelo Efitroarana, Sibien Mahereza, Enafa Jaonarisoa, Edouard Ramahatratra, Andry Randrianandrasana, Jeannicq Randrianarisoa and Jacky Youssouf, who have taught me so much over the years.

Several people have ploughed through and commented upon drafts of chapters or the whole manuscript, and I hope there are fewer mistakes and omissions now as a result. I thank you all: Chloe Chen-Kraus, Ronadh Cox, Brooke Crowley, Kristina Douglass, John Flynn, Laurie Godfrey, Steve Goodman, Julia Jones, David Krause, Christian Kull, Caroline Lehmann, Pierre-Yves Manguin, Sarah Osterhoudt, Chantal Radimilahy, Jean-Aimé Rakotoarisoa, Jeannin Ranaivonasy, Reinaldo Rasolondrainy, Joelisoa Ratsirarson, John Silander, Eleanor Sterling, Maria Vorontsova, Henry Wright and Anne Yoder.

Words are one thing, illustrations another, and I am extremely grateful to everyone who promptly, kindly and generously responded to anxious emails asking 'do you happen to have a brilliant photo of . . . ?' or 'please could you possibly draw . . . ?' Olivier Langrand opened new doors for me, inadvertently re-establishing a cherished connection with the Rajaonarison family, and I offer particular thanks to him.

If this book succeeds in interesting general readers, much of the credit goes to Wendy Strothman, my agent. She somehow managed to deliver harsh critiques without being terminally discouraging, and she never gave up on me (though she must have been tempted to). Through her, I met Andrew Gordon in London and, through him, Myles Archibald at William Collins, who inspired me to work further on an early draft. I am deeply grateful to Myles, and I thank Hazel Eriksson and the team who shepherded the book into print, particularly Sally Partington, the brilliantly zealous copy-editor, who won my heart by asking for more about vegetarian crocodiles. At the University of Chicago Press, Joe Calamia has been a wonderful editor and colleague, pushing me on many fronts without ever making me feel put-upon.

Clive Bush gave me my epigraph and shared writerly concerns and friendship along the way. Ben Gaskell taught me what I know about Malagasy quartz and steadfastly urged me on, even as he turned my ideas upside down and made me think again, over and over again. Enduring thanks to both of you.

It is my great good fortune to have amazing siblings. Finishing a book herself as I ploughed on with mine, Jean Baker offered advice, support, and a sisterly shoulder at low points, and my brother Angus Richard championed the cause of the general reader (I hope you find this a better read than earlier drafts.)

I have dedicated this book to our daughters Bessie and Charlotte, and their husbands Alex and Ed. They have been steadfast anchors throughout this book's long journey and during dark stretches in my life along the way. They have given me new and interesting things to think about, and love, laughter and fun. I am enormously proud of them and their children, my grandchildren – and grateful beyond words.

Around 2010, Bob and I decided to write a book together about Madagascar's long history, and we published an article that was effectively an outline for the book. After his death in 2013, I gave up all thought of writing, but then slowly began again. This is not the book we would have written together, but Bob's ideas thread their way through it, he himself was ever present as I wrote, and I could never have written it without him.

Alison Richard
Middle Haddam
October 2021

NOTES

Chapter 1

8   *a primate version of a giant panda*: Mittermeier et al. (2010, p. 579)

9   *Babokoto live in family groups*: This account draws on Pollock (1986), Powzyk & Thalmann (2003), and Giacoma et al. (2010).

9   *Far down below on the forest floor*: This account draws on Raxworthy (1991), Glaw & Vences (2007), and Glaw et al. (2021).

11   *how a distinguished ornithologist describes them*: Langrand (1990, p. 220–221)

11   *'Coy coy coy coy'*: Langrand (ibid) describes the call this way. For my part, I find it impossible to come up with words that actually sound like the call of a bird. I urge the reader to go and hear it for him- or herself.

11   *These iconic denizens of the forest*: A baobab fruit reached northwest Australia sometime between 17 and 7 million years ago, in a mighty and improbable sea voyage from Madagascar. For this and further baobab lore, see Baum (2003) and articles cited there.

12   *'Short, fat, gnarly branches'*: This phrase comes from Baum (ibid, p. 339).

13   *'go about life the way a chameleon walks'*: Raselimanana & Rakotomalala (2003, p. 967)

13   *'as for za, the Creator planted them upside down'*: Jolly (1980, p. 157)

15   *Our biological destiny is self-destruction*: Gray (2002)

15   *human ingenuity . . . will find a way through*: Browne (2019)

15   *giving rise to the evolution of new species as well as extinctions*: Thomas (2017)

15 *if only we try harder . . . all will not end in disaster*: Balmford (2012)

16 *This story . . . is flawed at best and plain wrong at worst*: Christian Kull (2000) writes cogently about this.

16 *The figure is used time and again*: e.g. Bradt et al. (1996)

17 *at least 27 articles . . . presenting it as a fact*: McConnell & Kull (2014)

18 *'Myths get made unbeknownst to man'*: Lévi-Strauss (1979, p. 17)

18 *the scents male ring-tailed lemurs . . . produce*: https://www.sciencefocus.com/news/lemurs-create-a-fruity-fragrance-to-attract-mates/

18 *A comedy film*: *Fierce Creatures*; Cleese & Johnstone (1997) wrote the script.

21 *Bob vastly expanded what I thought about*: His last article (Dewar, 2014) was in many ways the outline of a book we planned to write together.

Chapter 2

25 *The African coast disappeared from view*: The coastlines of Madagascar and Africa are not straight and so, depending on where you measure, the shortest distance is actually little over 400 km and the longest almost 1,000 km. Distances to India, Indonesia and Australia vary likewise.

25 *'Madagascar: heads it's a continent, tails it's an island'*: de Wit (2003)

27 *'if Gondwanaland existed, Madagascar was part of it'*: Brenon (1972, p. 88)

27 *the star of the continental drift saga*: de Queiroz (2014) describes the life and work of Alfred Wegener in detail.

29 *'in the right direction to fit drift theory'*: Wegener (1929, p. 32)

29 *an animated version*: The animation can be seen at http://www.reeves.nl/gondwana. For words, I draw on articles by de Wit (ibid), Wells (2003), Ali & Huber (2010), and Ali & Krause (2011).

29 *The supercontinent of Gondwana*: The literal meaning of Gondwana, a term first used for the continent in the nineteenth century, is 'Forest of the Gonds'. Derived from Sanskrit, it originally referred to a region in central India.

31 *New studies of the Mozambique Channel's floor*: Masters et al. (2021)

32 *The Indian monsoon system interacts*: This account draws on Boivin et al. (2013)

33 *Sixty-five million years ago*: Ali & Huber (ibid)

34 *a climate similar to modern conditions*: Krause (2003), Ohba et al. (2016)

34 *wide enough to swallow even a big island*: Wells (ibid)

34 *The island's relief also played a crucial role*: de Wit (ibid), de Wit & Anderson (2003), Ohba et al. (ibid)

36 *the grasslands of north Africa turned into the Sahara Desert*: Virah-Sawmy et al. (2010. p. 515)

37 *patterns of rainfall . . . are unpredictable*: Dewar & Wallis (1999), Dewar & Richard (2007)

38 *My* lambahoany *flapped wet*: A *lambahoany* is a length of cotton cloth. It serves many purposes, but is primarily worn like a sarong in many rural areas.

40 *the paths taken by cyclones in the Indian Ocean*: Mavumé et al. (2009)

41 *it still has a geological heartbeat*: Ronadh Cox gave me this lovely phrase.

42 *The land is 'terribly gullied'*: Wells & Andriamahaja (1993, p. 1)

42 *rests on scant or outdated evidence*: Jarosz (1993), Kull (2000)

42 *natural processes are also at work*: Wells et al. (1991), Wells & Andriamahaja (ibid), Cox et al. (2009, 2010)

45 *sediment levels in Madagascar's western rivers*: Kull (ibid, p. 437)

Chapter 3

50 *These trees belonged to an ancient plant group*: This group is called Gymnosperms. They carry seeds on the surfaces of their leaves or have leaves modified to form cones.

50 *Their similarity to the trees of Northern Pakistan at the time*: Hankel (1993)

50 *It was the third of five waves*: Wake & Vredenburg (2008)

51 *a whole new world of vegetation*: Wright & Askin (1987)

52 *kinds unlike any seen anywhere in the world today*: Flynn & Wyss (2002) provide readers seriously enchanted by strange creatures with a starting point to explore the team's discoveries in more detail.

53 *boundaries that often do not . . . exist in nature*: James Prosek (2020) writes lyrically about this.

54 *the discovery of a fifth kind of animal*: Kammerer et al. (2020)

56 *That killer was likely drought*: Rogers & Krause (2007)

56 *The reptiles, amphibians and fish . . . were very different*: Krause et al. (1997)

57 *the only theropods with demonstrated cannibal tendencies*: Rogers et al. (2007)

57 *perhaps one species of flying dinosaur*: O'Connor & Forster (2010)

58 *Known from a well-preserved fossil skull*: Buckley et al. (2000)

58 *Only one species has been identified and described so far*: Evans et al. (2008)

58 *These carnivorous frogs . . . have been dubbed 'hopping heads'*: Lappin et al. (2017)

59 *identified as a marsupial mammal*: Krause (2001)

59 *A skull discovered in the Mahajanga Basin*: Krause et al. (2014)

60 *And then there's* Andalatherium hui: Krause et al. (2020)

61 *before Madagascar was surrounded by water*: Ali & Krause (2011)

62 *the increasing isolation of island life*: Samonds et al. (2013)

62 *The fifth global wave of extinctions followed*: Jablonski (1995)

Chapter 4

67 *'a ghost lineage'*: Norell (1992)

71 *'Any event that is not absolutely impossible . . . becomes probable'*: Simpson (1952, p. 174)

72 *less than four weeks for a mat to drift across the Channel*: Ali & Huber (2010)

72 *Some, notably chameleons*: Rafferty & Reina (2012)

73 *Voyages across the Atlantic Ocean*: de Queiroz (2014)

73 *The ancestors of Madagascar's living mammals*: Nowack & Dausmann (2015), Blanco et al. (2018)

73 *'snoozing through disaster'*: Anne Yoder conjured this vision during a lecture at Yale in May 2016.

74 *All but two species of living frogs*: My primary frog sources are Vences

et al. (2004) and Vences & Raselimanana (2018); Vieites et al. (2009) explain the idea of candidate species; Glaw & Vences (2007) offer an excellent field guide (though now a bit out of date) to both amphibians and reptiles.

75 *With 13 families today*: Glaw & Raselimanana (2018)

75 *lizards certainly got around*: Raxworthy et al. (2002) and Tolley et al. (2013) are my main sources on lizards.

76 *Land tortoises . . . spread widely*: For tortoises, I draw on Caccone et al. (1999), Palkovacs et al. (2002), Gerlach et al. (2006), and Wilmé et al. (2016)

78 *Madagascar's crocodiles*: My crocodile sources are Brochu (2007), Meador et al. (2019), and Hekkala et al. (2021)

78 *Most Malagasy snakes*: For snakes, I relied on Noonan & Chippendale (2006), Crottini et al. (2012), and Samonds et al. (2013).

79 *First to arrive were lemurs*: For living lemurs, I draw on Wright (1999), Yoder & Yang (2004), Kappeler & Schaffler (2008), Ganzhorn et al. (2009), Yamashita et al. (2010), Mittermeier et al. (2014), Gunnell et al. (2018), Tattersall & Cuozzo (2018), and Schubler et al. (2020)

81 *Sub-fossil remains are the only evidence of 17 lemur species*: This brief introduction is based on Richard & Dewar (1991) and Godfrey & Jungers (2003).

83 *After lemurs came tenrecs*: Eisenberg & Gould (1970), Goodman & Benstead, eds. (2003), and Goodman et al. (2018) are my tenrec sources.

83 *can be mistaken for dead*: The late Martin Nicoll – tenrec-ologist and fount of knowledge about all Madagascar's wildlife – actually made this mistake, or so he told me long ago.

83 *The next ancestor . . . to arrive*: This account draws on Yoder et al. (2003), Goodman (2012), and Meador et al. (ibid). The order's name is confusing, because many animals eat flesh and are 'carnivorous', without being members of the order Carnivora; those belonging to the order all trace back to a common ancestor, and members of the order in Madagascar are accorded a separate family.

84 *For me, the most intriguing of the 28 endemic rodent species*: My jumping

giant rat sources are Sommer (2003) and Goodman & Soarimalala (2018)

84 *There were probably three species*: It is not completely certain there were three species rather than two, one of which had members that varied widely in size. Assuming three species, their scientific names are *Hippopotamus lemerlei* (in the west), *Hippopotamus laloumena* (in the east), and *Hexaprotodon madagascariensis* (in the central highlands) (Rakotovao et al. 2014, Stuart 2021)

85 *they have an order all to themselves: Bibymalagasia*: MacPhee (1994) and Buckley (2013) recount the history of ideas about this mysterious animal.

85 *the world's major bat lineages*: For bats, I turned to Teeling et al. (2005), Goodman (2011), and Goodman & Ramasindrazana (2018)

86 *Turning to birds, the puzzle persists*: For birds, living and extinct, my sources are Langrand (1990), Hawkins & Goodman (2003), Yoder & Nowak (2006), Goodman & Jungers (2014), Yonozawa et al. (2017), Goodman & Raherilalao (2018), and Hansford & Turvey (2018).

89 *their energy turnover is very low indeed*: Richard & Nicoll (1987)

90 *It is also a saga of connection and isolation*: Samonds et al. (ibid)

Chapter 5

95 *the antiquity of grasslands in Madagascar*: Bond et al. (2008)

95 *Ancient plant (and animal) remains . . . to infer past climates*: In a new variant of this approach, geochemical analyses of coral reefs have been used to infer rainfall variability along Madagascar's coastline, but only on a timescale in the hundreds or thousands of years thus far (Grove et al. 2013).

96 *which occur nowhere else in the world*: Lowry et al. (2018)

97 *The best overview of Madagascar's vegetation today*: Moat & Smith (2007)

99 *groups according to their likely land of origin:* Perrier de la Bâthie (1936)

100 *Not surprisingly, the closest relationships of all*: Schatz (1996), and Buerki et al. (2013)

100 *only one . . . old enough to predate the break-up of Gondwana*: These small

evergreen trees or shrubs are called *Takhtajania perrieri*. The genus has only one species, found in small patches of moist forest in eastern Madagascar (Thien et al. 2003).

100 *Take* Canarium, *for example*: Federman et al. (2018)

101 *pumice stone thrown up by . . . Krakatoa*: Koechlin et al. (1974)

101 *When Madagascar began its arduous passage*: Wells (2003)

101 *Molecular evidence suggests*: Buerki et al. (ibid)

103 *A few decades back, a distinguished botanist concluded*: Koechlin et al. (ibid)

103 *Well-studied grasslands . . . provide a roadmap*: Goodman & Jungers (2014)

103 *Grasslands is probably the term*: White (1983)

104 *the models ignore the roles played by other forces*: Bond & Keeley (2005), and Lehmann et al. (2014)

104 *minor changes in climate and fire regime have shifted the landscape*: Ekblom & Gillson (2010)

104 *virtually treeless in a few decades*: Western & Maitumi (2004)

104 *this proposition, called succession theory*: Clements (1916)

105 *An isotopic signal showing a mixture of C3 and C4*: Jacobs et al. (1999)

105 *The conditions enabling C3 and C4 grasses*: Strömberg (2011)

107 *'biologists should take a fresh look at Madagascar's grasslands'*: Bond et al. (ibid, p. 1753)

109 *almost all modern endemic grasses*: A mysterious bamboo-like C3 grass is a single exception, probably arriving over 20 million years ago (Besnard et al. 2013).

109 *there is still much to learn*: Joseph & Seymour (2020)

109 *burning them annually to encourage new growth*: Kull (2004) gives an in-depth account of natural and managed fire regimes in Madagascar.

109 *Maria and her colleagues set out to answer this question*: Vorontsova et al. (2016)

110 *many endemic Malagasy plants have similar abilities*: Koechlin et al. (ibid)

110 *intriguing new evidence from the central highlands*: Solofondranohatra et al. (2020)

111 *for the last 150,000 years or so*: Gasse & Van Campo (2001); there

are no radiocarbon dates for older sections of the core, and 150,000 years is a tentative date inferred on other grounds.

111 *the record for the last 11,000 years*: Burney (1987a, b, c)

111 *yet another dimension*: Samonds et al. (2019)

115 *Under this scenario*: Wilmé et al. (2006)

116 *studied in detail so far*: Sources of these studies can be found in Blair et al. (2015).

116 *pipe up the mouse lemurs*: Yang & Yoder (2003)

116 *a 'telltale species' for the presence of grasslands*: Yoder et al. (2016, p. 8054). Quémeré et al. (2012) found a similar genetic signal dating back 5,000–10,000 years in golden-crowned *sifaka* living in fragmented habitats in the northeast.

117 *Evolutionary processes are still at work today*: Heckman et al. (2006), Rasoazanabary (2011), and Agostini et al. (2017). For another example, a nicely titled article recently described three populations of the tenrec species *Oryzorictes hova* on their way to becoming separate species (Everson et al. 2018).

Chapter 6

121 *'extreme' life history patterns are also common*: Stearns (1992), and Flannery (1994)

121 Tenrec ecaudatus *has the biggest known litters*: Nicoll (2003)

122 *A species of chameleon found in the southwest*: Karsten et al. (2008)

122 *some of the 'slowest' animals on Earth*: Dewar & Richard (2007)

123 *the extinct lemurs matured and reproduced more slowly*: Godfrey & Rasoazanabary (2012)

123 *the curious case of the black-and-white ruffed lemur community*: Baden et al. (2013)

125 *An aye-aye's body looks as if*: The spare parts image is from Bomford (1981), but most of what is known about wild aye-ayes comes from a landmark study by Eleanor Sterling (1993), and the description of them licking larvae like ice-cream cones is hers.

126 *Yet animals living in captivity find cavities*: Erickson et al. (1998)

126 *some lemurs still play . . . an ancient primate role*: Kress et al. (1994)

127 *The living lemurs . . . as seed dispersers*: Wright et al. (2011), and Razafindratsima et al. (2014)

127 *the likelihood that the 'orphaned' tree species will survive*: Albert-Daviaud et al. (2020)

128 *grassland evolution and the evolution of grassland faunas*: Strömberg (2011)

129 *But recent re-analysis of these finds*: Goodman & Jungers (2014)

129 *candidates for an ancient grassland community*: Bond et al. (2008)

130 *Animals are not quite what they eat*: This observation comes from the title of Codron et al. (2011).

131 *Most of the 17 or so extinct species*: Godfrey et al. (1997)

131 *The* coup de grace *for most giant lemurs*: Godfrey & Jungers (2003)

132 *Monkey lemurs were different*: Godfrey et al. (2015)

132 *Recent isotope analysis of a sub-fossil bone*: Samonds et al. (2019)

133 *'would do credit to a golf course'*: Eltringham (1999, p. 78) made the golfing comparison.

133 *Isotope analysis of a sub-fossil hippo bone*: Samonds et al. (ibid)

134 *the remains from two sites signal a mix*: Godfrey & Crowley (2016)

134 *A plausible scenario*: My account of the Aldabra population is based on Merton et al. (1976), Coe et al. (1979), and Gibson & Hamilton (1983).

135 *evidence assembled by the tortoise-watchers of Aldabra*: In an experiment on an off-shore Mauritian nature reserve, Aldabran tortoises and radiated tortoises from Madagascar quickly cleared woody vegetation and established grazing lawns in the places they were introduced (Griffiths et al. 2013), and giant tortoises in the Galapagos Islands exhibit many parallels too (Gibbs et al. 2010).

135 *Madagascar's extinct elephant birds*: My sources are Clarke et al. (2006), Tovondrafale et al. (2014) and Hansford & Turvey (2018, in press) for elephant birds, Milton et al. (1994) for ostriches, and Bond & Silander (2007) for vegetation.

136 *As many as four elephant bird species*: They may also have been in the east, but that cannot be known until sub-fossil sites are discovered in this region (Goodman & Jungers ibid).

138 *The biomass of termites*: For termite lore, I draw on Aanen & Eggleton (2005), Nobre et al. (2010), and Okullo & Moe (2012)

140 *some ant species are . . . important contributors*: Fisher & Robertson (2002)
140 *bird species that do well in open country*: Langrand (1990)
140 *Miombo woodlands . . . as a model*: Goodman & Jungers (ibid, p. 30)
141 *A single scenario also misses the complexity*: Crowley et al. (2021)
141 *a rapid cascade of consequences*: Ekblom & Gillson (2010), and Carroll (2016)
141 *Instead of assuming stability*: van der Leeuw (2000)

Chapter 7

145 *the Common Era:* The start of the Common Era (CE) corresponds in time with *Anno Domini* (AD).

146 *The Malagasy historian Raombana*: sent to England for eight years with his twin brother for their education in the early nineteenth century, Raombana returned to Madagascar, played a central role in state affairs, and wrote extensively about the island's history – in English. In 1980, a complete set of his known work was published, edited, and translated into French by Simon Ayache.

148 *Signs of their presence*: Hansford et al. (2018)

150 *The reliability of age estimates*: Douglass et al. (2019)

150 *Cut-marks on an assortment of . . . bones*: Godfrey et al. (2019)

151 *The earliest glimpse of a place where people camped*: Dewar et al. (2013)

151 *the balance of evidence supports . . . early date*: Ekblom et al. (2016)

154 *Small forager groups began camping*: Douglass & Zinke (2015)

154 *A few linguistic clues*: Dewar (2014)

156 *Consider two of them*: See Wright & Rakotoarisoa (in press) for a comprehensive review of the archaeological record, including these sites.

157 *Trade linked Madagascar's settlers*: Mack (2007), Boivin et al. (2013), Campbell (2016), and Beaujard (2019)

158 *A first-century handbook of the Indian Ocean*: Casson (1989)

158 *The voyage of the* Sarimanok: Hobman & Burley (1989)

159 *genetic make-up of the Malagasy people*: Pierron et al. (2017)

160 *Efforts to locate more precisely . . . within Indonesia*: Dahl (1951), Adelaar (2009), and Kusuma et al. (2016)

160   *These boats transported great loads*: Manguin (2016)
160   *Settlers could have travelled*: Kusuma et al. (2015)
161   *Words for domestic animals are more closely linked to Africa*: Beaujard (ibid), and Blench (2008)
161   *Botanical samples . . . are strikingly different*: Crowther et al. (2016)
162   *human activities may have caused some of these effects*: Campbell (ibid)
162   *Political developments in Southeast Asia*: Beaujard (ibid)
163   *different maternal and paternal genetic lineages*: Pierron et al. (ibid)
163   *With the age of arrivals over*: Wright & Rakotoarisoa (ibid)
163   *people were clearing and planting the land*: Burney (1987c)
164   *Andranosoa sits on the bank of a river*: Rasamuel (1984), Parker Pearson (2010)
165   *Cattle were introduced more than once*: Fuller & Boivin (2009)
165   *the excavation of this important site*: Radimilahy (1998)
166   *may be the vestige of a well-worn route*: Vérin (1986, p. 14)
166   *exquisite pieces crafted in the Islamic world*: Horton et al. (2017)
167   *written records [of] small, dynastic kingdoms 'too numerous to count'*: Alfred Grandidier and his son Guillaume compiled texts written between 1550 and 1668, commissioned translations into French from Dutch, English, German, Italian, Spanish and Latin, and published them in nine volumes in 1904. The quotation is from an early seventeenth-century report by a Goan Jesuit (Vol 2: p. 5).
167   *key to Merina prosperity*: Campbell (2005)
168   *Slavery has a long history in Madagascar*: Campbell (ibid), Larson (2000), Graeber (2007)
168   *The travels of an unnamed Malagasy ex-slave*: Larson (2009)
168   *close to 3.3 million people*: Campbell (ibid)
169   *Overseas origins are not part of Malagasy identity today*: Mack (ibid), quote from Rakotoarisoa (2002, p. 25)
169   *the experience of Sarah Osterhoudt*: Osterhoudt (2017, and pers. com.)
171   *a linguistic chequerboard of Malagasy dialects and Bantu languages*: Vérin (1980)
171   *In the early nineteenth century, Nicholas Mayeur*: Larson (ibid, p. 35)
172   *'Honeysuckle Rose'*: Kristina Douglass told me this.

Chapter 8

176 *Its diverse regions have been likened to*: Martin (1972, p. 371)

176 *Each small island has a long and distinct history*: Wright & Rakotoarisoa (1997)

179 *Climate fluctuations and their impact*: Mahé & Sourdat (1972), Burney (1993), Vallet-Coulomb et al. (2006), and Faina et al. (2021)

179 *Water birds disappeared too*: Goodman et al. (2013)

179 *easier to live by fishing or trading than farming*: Dewar (1997)

179 *Sediment cores from four sites:* Virah-Sawmy et al. (2009)

180 *evidence of fishing and cattle husbandry*: Wright & Rakotoarisoa (in press)

181 *Interacting effects of rainfall, soil and grazing pressure*: Ratovonamana et al. (2013)

182 *Some have questioned its authenticity*: Jolly (2004, p. 80). Parker Pearson (2010) likewise concludes that the account is based on real events.

182 *'the landscape does not seem all that different . . .'*: Jolly (ibid, p. 92)

183 *Feral cattle . . . still roam forests today*: Goodman et al. (2003)

183 *the feral population extended the human footprint*: Dewar (1984), and Burney et al. (2003)

183 *I am less convinced now that he was right*: If he were, there should be less forest in the south than there is. Speaking with Charlie Burrell – who with Isabella Tree launched the rewilding initiative at Knepp Castle in Sussex – he was emphatic that free-roaming cattle in English woodlands consume small saplings but generally do no harm to mature trees and little or none to young ones.

185 *expanded into a mosaic of forest, woodland and grassland*: Burney (1987 a, b, c), Gasse & Van Campo (1998), and Samonds et al. (2019)

186 *The History of Kings*: Larson (1995) points out that these stories were told by noblemen and royal courtiers, and other voices and perspectives were not recorded.

186 *brought fleeting prosperity to the region*: This description draws on detailed accounts by Berg (1981) and Campbell (2013).

186 *Charcoal production for ironworking*: Henry Wright told me this.

188 *The northwest has its own distinctive climate history*: Wang et al. (2019)

188   *Five stalagmites from Anjohibe and Anjokipoty*: Burns et al. (2016), and Wang et al. (ibid)

189   *Pollen from a sediment core . . . archaeology another dimension still*: Matsumoto & Burney (1994), and Wright et al. (1996)

191   *It seems likely that forests returned*: Wright & Rakotoarisoa (ibid)

192   *signals of a change in human economy*: Dewar et al. (2013)

192   *a 'subsistence shift hyphothesis'*: Godfrey et al. (2019)

193   *seem to have been least where people settled earliest*: Dewar (1997)

193   *according to historical accounts*: Jarosz (1993), and Campbell (ibid)

194   *Elephant bird imagined*: H.G. Wells published a novella called *Aepyornis Island* in 1894, but the cover of *Pearson's magazine* featuring it in 1905 is more to my liking than the novella's original cover.

Chapter 9

195   *a creature they called kilopilopitsofy*: Burney & Ramilisonina (1998)

195   *a written account four centuries earlier*: Flacourt (1658)

196   *a large number of smaller mammal species*: Muldoon et al. (2009)

197   *A broad chronology of what happened thereafter*: Crowley (2010)

198   *all dated within a 1,500-year period*: Godfrey et al. (2019)

198   *further evidence of the decline of the mega-fauna*: Burney et al. (2003), and Raper & Bush (2009)

201   *the best evidence so far that early settlers hunted*: Goodman & Jungers (2014) review the sparse and often problematic evidence from other localities for the hunting of lemurs and hippopotamuses.

201   *no indication of a bottleneck*: Lawler (2008)

202   *A new discovery of prehistoric cave art*: Burney et al. (2020)

202   *Until then, the only well-described cave art*: Rasolondrainy (2012)

203   *'consistent with that of the extinct sloth lemurs'*: Burney et al. (ibid, p. 12)

204   *the most easily found trace*: Dewar (1984)

204   *Dense concentrations of broken eggshell*: Parker Pearson (2010)

205   *Piles of elephant bird eggshell*: Parker Pearson (ibid), Douglass (2016)

205   *according to two nineteenth-century accounts*: Battistini et al. (1963), and Parker Pearson (ibid)

205 *beads were important exchange items*: Mitchell (1996)

205 *Velondriake is still the only place*: Douglass (ibid)

206 *eggs laid . . . during the time the site was occupied*: Parker Pearson (ibid, p. 88)

206 *Marco Polo wrote a fantastical account*: Parker Pearson (ibid)

207 *the journal of a Frenchman*: Translation by Ross Barnett (https://twilightbeasts.org/2014/12/04/the-most-lonely-places/)

207 *a reputation for ferocity, and a 'malignant eye'*: Low (2016)

207 *A blood-chilling report of a cassowary attack*: Flannery (2017)

208 *A computer simulation of human settlers*: Alroy (2001)

209 *Giant lemurs and tortoises matured and reproduced slowly*: Godfrey et al. (1997)

210 *a synergy of effects that differed regionally*: Dewar (ibid), Burney et al. (ibid), Goodman & Jungers (ibid), and Hixon et al. (2021)

210 *the few reliable dates*: Crowley & Samonds (2013)

211 *'an ecological catastrophe that was too gradual to be perceived'*: Alroy (ibid, p. 1896)

Chapter 10

214 *struggles for control of the land*: Jarosz (1993), and Evers et al. (2013)

215 *often on the losing side*: Poudyai et al. (2018)

215 *rosewoods are under imminent threat*: Barrett et al. (2010), and Randriamalala & Liu (2010)

216 *Creation of the colonial state in 1896*: Deschamps (1972), and Evers et al. (ibid, p. 8)

217 *a system of shifting cultivation*: Scales (2014)

217 *3 to 7 million hectares lost*: Jarosz (ibid)

217 *The wide range of figures*: McConnell & Kull (2014)

217 *Twelve nature reserves were established*: For a detailed review of the history and current status of terrestrial protected areas, see Goodman et al., eds. (2018).

218 *'a messy and difficult task'*: McConnell & Kull (ibid, p. 67)

218 *Those left are increasingly fragmented*: Vieilledent et al. (2018)

218 *One commonly cited study*: Harper et al. (2007)

218 *Boundaries have . . . even expanded*: Elmqvist et al. (2007)

219 *different lemur species divide up forest resources*: Sussman (1974)

220 *According to a recent study*: Albert-Daviaud et al. (2018)

220 *Hunting to put food on the table*: Randrianandrianina et al. (2010), Reuter et al. (2016)

220 *Newcomers may not respect prohibitions*: Jones et al. (2008)

221 *a recent study enquired of villagers*: Razafimanahaka et al. (2012)

222 *56 reptile and 16 amphibian species are targeted*: Raselimanana (2003)

222 *people have brought new species*: Kull et al. (2014)

222 *They covered the floors*: Actman (2018)

224 *Then came a maize boom*: Scales (ibid) describes the arc of this boom in depth.

225 *By 2009, there were foreign investments*: Evers et al. (ibid, p. 2)

225 *demand for charcoal . . . drives another whole supply chain*: Gardner et al. (2016)

226 *conservation efforts really took off in the mid-1980s*: Kull (2014), Corson (2016), and Jolly (2015)

226 *coined the . . . phrase 'biodiversity hotspot'*: Myers (1988)

226 *a 'green grab' by foreigners*: Fairhead et al. (2012) coined the term 'green grab'.

226 *The reality is more complicated*: Evers et al. (ibid, p. 5)

227 *cultivators were by now identified as villains*: Jarosz (ibid, p. 372)

227 *an enduring symbol of protest against the state*: Kull (2004), Jarosz (ibid)

227 *a stream of policy developments*: This discussion draws on the evidence and insights of many researchers; references to work not cited individually can be found in a new review by Jones et al. (in press).

230 *One jaundiced view*: Pollini et al. (2014, pp. 187–188)

230 *a tsunami of extinctions'*: This vivid metaphor is from Lovejoy (2013).

230 *Some liken our species to a weed*: The idea of animals as weeds has long interested me (Richard et al. 1989); Quammen (2008) extended my thinking to include our own species.

231 *if one listens to farmers in Madagascar*: Kull et al. (ibid)

Chapter 11

234 *an international centre for training and research*: Wright & Andriamihaja (2002)

234 *A hundred and twenty-two protected areas*: Goodman et al., eds. (2018)

234 *'a dizzying range of forest types . . .'*: McConnell & Sweeney (2005, p. 223)

235 *the network is effective in reducing deforestation*: Eklund et al. (2016)

235 *The presence of people*: Osterhoudt (2021) expands this observation into a broader critique of the widespread propensity to dwell on situations where things do not work well.

235 *Marine environments are easier to protect*: Gardner et al. (2013)

235 *A clear demonstration of this*: Harris (2007)

236 *In 2004, working with Blue Ventures*: Several international and regional NGOs work with coastal communities today.

236 *let us call them the western lakes*: This account draws on Watson & Rabarisoa (2000), Watson et al. (2007), and Watson (2018).

239 *under the Ramsar Convention*: This is an intergovernmental treaty providing a framework for national action and international cooperation for the conservation and wise use of wetlands and their resources (ramsar.org).

241 *long history of a partnership forged*: Richard & Dewar (2001), Richard & Ratsirarson (2013), and Ranaivonasy et al., eds. (2016)

244 *protected forests . . . remain largely intact*: Ranaivonasy et al. (2016)

244 *I return to the village now*: My account draws on Osterhoudt (2017, 2020), and Dove's foreword to her book (2017).

245 *the regional port of Mananara*: Ellis (2007)

247 *'messy landscapes . . . make for grumpy ancestors'*: Osterhoudt (2017, p. 13)

248 *stewardship of tany is deeply embedded in Malagasy culture*: Osterhoudt (ibid, p. 113), and Kaufmann (2014)

248 *The success of . . . grassroots conservation efforts*: Sterling et al. (2017)

249 *Conservation dina work in some communities*: Rabesahala Horning (2003)

249 *the hijacking of a form of social contract*: e.g. Pollini et al. (2014)

249 *several things need to happen*: Jones et al. (2019)

250 *locally nurtured agroforestry systems*: Osterhoudt (ibid, p. 5)

250 *Concerned about the inequity*: Dove (p. *x* in Osterhoudt ibid)

Chapter 12

254 *'Then they left, and now they are coming back'*: Keller (2015, p. 202)

254 *It was a wrenching role*: Eva herself wrote that, looking at the park from the perspectives of the local community and the international funding organisation, it was as if they 'were engaged in two completely different stories, following different plots and starring different protagonists' (ibid, p. 216).

255 *The French government declared a protectorate*: This account draws on Ellis (1985) and Campbell (2005).

257 *'settler as well as scientist'*: Middleton (1999, p. 228)

257 *extolled in an adulatory obituary*: Humbert (1958)

257 *one turns to his monumental work*: Perrier de la Bâthie (1921)

258 *a hefty tome on Magagascar's geography*: Gautier (1902)

258 *Not so, countered Perrier de la Bâthie magisterially*: Perrier de la Bâthie (ibid, pp. 173–174)

258 *Alfred Grandidier . . . remarked around the same time*: Grandidier (1898, p. 84)

258 *people introduced fire to the island in his view*: I have found only one passing reference by him to fire occurring naturally (Perrier de la Bâthie ibid, p. 171).

258 *But people also burned 'by simple habit . . .'*: Perrier de la Bâthie (ibid, p. 265)

259 *The Cochineal Insect Affair*: Middleton (ibid) provides a detailed account of this episode, from which come the quotations I use, translated from French by her.

261 *three different estimates*: His estimates in *La Végétation Malgache* (1921) are variously 9/10, 7/8 and 5/6 (pp. 3, 62, 262). The estimate shifts to 7/10 in *Biogéographie des Plantes de Madagascar* (1936, p. 11).

261 *evidence to counter it*: Notably Kull (2000)

261   *Myths are the way people 'reach by the shortest possible means . . .'*: Lévi-Strauss (1979, p. 17)

261   *The myths of Paradise and the Garden of Eden*: Delumeau (1995)

262   *'the veritable promised land for naturalists'*: quoted by Feeley-Harnik (2001)

262   *'The Site of Paradise'*: from Haeckel (1876)

262   *But, buried within their texts*: Feeley-Harnik (ibid)

262   *the evocative power of his prose*: The quotes here are from Perrier de la Bâthie (1921, pp. 7, 44, 99, and 1936, p. 3), translated from French by the author. Language of this kind echoes through both volumes, however.

264   *recent popular representations*: Such invocations are actually quite common around the world, especially applied to islands (Grove 1995). What singles out Madagascar is their continuing hold on public perceptions.

264   *images conjured in the* National Geographic *magazine*: These are from Marden (1967), Jolly (1987), and Shea (2009), respectively.

266   *at just one site*: Bemanevika, in the northwest.

266   *Many ecological processes play out*: Willis et al. (2010)

267   *new ways of approaching conservation in the southwest*: Virah-Sawmy et al. (2016)

268   *A better description is of countryside transformed*: See Richard & O'Connor (1997), Kull (2004), and Pollini (2009) for more on the line between degradation and transformation, and Tsing (2015) for a broader discussion.

268   *Such dichotomies are still widely used:* Bond & Parr (2010)

268   *The abundance of trees in the region*: McConnell et al. (2015), and Bond (2016)

269   *Their value . . . is low, though not zero*: Kull et al. (2014)

269   *several different Malagasy words used regionally*: Scales (2014, p. 106)

270   *a distinction between self-professed authorities . . . and the tellers of* tantara: Osterhoudt (2017)

271   *'For landscapes to be truly sustainable'*: Osterhoudt (ibid, p. 12)

BIBLIOGRAPHY

Aanen, D. K., & Eggleton, P. (2005). Fungus-growing termites originated in African rain forest. *Current Biology, 15*(9), 851–855.

Actman, J. (April 20, 2018). Stench leads to home crawling with stolen tortoises – 10,000 of them. From https://news.nationalgeographic.com/2018/04/wildlife-watch-radiated-tortoises-poached-madagascar/

Adelaar, A. (2009). Towards an Integrated Theory About the Indonesian Migrations to Madagascar. In P. Peregrine, I. Peiros, & M. Feldman (Eds.), *Ancient Human Migrations: A Multidisciplinary Approach* (pp. 149–172). Salt Lake City, UT: University of Utah.

Agostini, G., Rasoazanabary, E., & Godfrey, L. R. (2017). The befuddling nature of mouse lemur hands and feet at Bezà Mahafaly, SW Madagascar. *American Journal of Primatology, 79*(9), e22680.

Albert-Daviaud, A., Buerki, S., Onjalalaina, G. E., Perillo, S., Rabarijaona, R., Razafindratsima, O. H., Sato, H., Valenta, K., Wright, P. C., & Stuppy, W. (2020). The ghost fruits of Madagascar: Identifying dysfunctional seed dispersal in Madagascar's endemic flora. *Biological Conservation, 242*, 108438.

Albert-Daviaud, A., Perillo, S., & Stuppy, W. (2018). Seed dispersal syndromes in the Madagascan flora: The unusual importance of primates. *Oryx*, 1–9.

Ali, J. R., & Huber, M. (2010). Mammalian biodiversity on Madagascar controlled by ocean currents. *Nature, 463*(7281), 653–656. doi: 10.1038/nature08706

Ali, J. R., & Krause, D. W. (2011). Late Cretaceous bioconnections between Indo-Madagascar and Antarctica: Refutation of the Gunnerus Ridge causeway hypothesis. *Journal of Biogeography, 38*(10), 1855–1872. doi: 10.1111/j.1365-2699.2011.02546.x

Alroy, J. (2001). A multispecies overkill simulation of the end-Pleistocene megafaunal mass extinction. *Science, 292*(5523), 1893–1896.

Baden, A. L., Wright, P. C., Louis, E. E., & Bradley, B. J. (2013). Communal nesting, kinship, and maternal success in a social primate. *Behavioral Ecology and Sociobiology, 67*(12), 1939–1950.

Balmford, A. (2012). *Wild Hope: On the Front Lines of Conservation Success.* Chicago: University of Chicago Press.

Barrett, M. A., Brown, J. L., Morikawa, M. K., Labat, J-N., & Yoder, A. D. (2010). CITES designation for endangered rosewood in Madagascar. *Science, 328*(5982), 1109–1110.

Battistini, R., Vérin, P., & Rason, R. (1963). Le site archéologique de Talaky: cadre géographique et géologique; premiers travaux de fouilles; notes ethnographiques sur le village actuel proche du site. *Annales Malgaches, Faculté des Lettres et Sciences Humaine, Antananarivo,* 112–153.

Baum, D. (2003). Bombacaceae, Adansonia, Baobab, Bozy, Fony, Renala, Ringy, Za. In S. Goodman & J. Benstead (Eds.), *The Natural History of Madagascar* (pp. 339–342). Chicago: University of Chicago Press.

Beaujard, P. (2019). *The Worlds of the Indian Ocean: Volumes 1 and 2.* Cambridge, UK: Cambridge University Press.

Berg, G. M. (1981). Riziculture and the founding of monarchy in Imerina. *The Journal of African History, 22*(3), 289–308.

Besnard, G., Christin, P-A., Malé, P-J. G., Coissac, E., Ralimanana, H., & Vorontsova, M. S. (2013). Phylogenomics and taxonomy of *Lecomtelleae* (Poaceae), an isolated panicoid lineage from Madagascar. *Annals of Botany, 112*(6), 1057–1066.

Blair, C., Noonan, B., Brown, J., Raselimanana, A., Vences, M., & Yoder, A. (2015). Multilocus phylogenetic and geospatial analyses illuminate diversification patterns and the biogeographic history of Malagasy endemic plated lizards (Gerrhosauridae: Zonosaurinae). *Journal of Evolutionary Biology, 28*(2), 481–492.

Blanco, M. B., Dausmann, K. H., Faherty, S. L., & Yoder, A. D. (2018). Tropical heterothermy is 'cool': The expression of daily torpor and hibernation in primates. *Evolutionary Anthropology: Issues, News, and Reviews, 27*(4), 147–161.

Blench, R. (2008). The Austronesians in Madagascar and their interaction with the Bantu of the East African coast: Surveying the linguistic evidence for domestic and translocated animals. *Studies in Philippine Languages and Cultures, 18*, 18–43.

Boivin, N., Crowther, A., Helm, R., & Fuller, D. Q. (2013). East Africa and Madagascar in the Indian Ocean world. *Journal of World Prehistory, 26*(3), 213–281. doi: 10.1007/s10963-013-9067-4

Bomford, E. (1981). On the road to Nosy Mangabe. *International Wildlife, 11*(1), 20–24.

Bond, W. J. (2016). Ancient grasslands at risk. *Science, 351*(6269), 120–122. doi: 10.1126/science.aad5132

Bond, W. J., & Keeley, J. E. (2005). Fire as a global 'herbivore': The ecology and evolution of flammable ecosystems. *Trends in Ecology & Evolution, 20*(7), 387–394.

Bond, W. J., & Parr, C. L. (2010). Beyond the forest edge: Ecology, diversity and conservation of the grassy biomes. *Biological Conservation, 143*(10), 2395–2404.

Bond, W. J., & Silander, J. A. (2007). Springs and wire plants: Anachronistic defences against Madagascar's extinct elephant birds. *Proceedings of the Royal Society of London B: Biological Sciences, 274*(1621), 1985–1992.

Bond, W. J., Silander, J. A., Ranaivonasy, J., & Ratsirarson, J. (2008). The antiquity of Madagascar's grasslands and the rise of C(4) grassy biomes. *Journal of Biogeography, 35*(10), 1743–1758. doi: 10.1111/j.1365-2699.2008.01923.x

Bradt, H., Schuurman, D., & Garbutt, N. (1996). *Madagascar Wildlife: A Visitor's Guide*. Chalfont St. Peter: Bradt Publications.

Brenon, P. (1972). The Geology of Madagascar. In R. Battistini &

G. Richard-Vindard (Eds.), *Biogeography and Ecology in Madagascar* (pp. 27–86). The Hague: Dr. W. Junk B.V.

Brochu, C. A. (2007). Morphology, relationships, and biogeographical significance of an extinct horned crocodile (Crocodylia, Crocodylidae) from the Quaternary of Madagascar. *Zoological Journal of the Linnean Society, 150*(4), 835–863.

Browne, J. (2019). *Make, Think, Imagine: Engineering the Future of Civilization.* New York: Pegasus Books.

Buckley, G. A., Brochu, C. A., Krause, D. W., & Pol, D. (2000). A pug-nosed crocodyliform from the Late Cretaceous of Madagascar. *Nature, 405*(6789), 941–944.

Buckley, M. (2013). A molecular phylogeny of *Plesiorycteropus* reassigns the extinct mammalian order 'Bibymalagasia'. *PLoS One, 8*(3), e59614.

Buerki, S., Devey, D. S., Callmander, M. W., Phillipson, P. B., & Forest, F. (2013). Spatio-temporal history of the endemic genera of Madagascar. *Botanical Journal of the Linnean Society, 171*(2), 304–329. doi: 10.1111/boj.12008

Burney, D. A. (1987a). Pre-settlement vegetation changes at Lake Tritrivakely, Madagascar. *Palaeoecology of Africa, 18,* 357–381.

Burney, D. A. (1987b). Late Holocene vegetational change in central Madagascar. *Quaternary Research, 28*(1), 130–143.

Burney, D. A. (1987c). Late Quaternary stratigraphic charcoal records from Madagascar. *Quaternary Research, 28*(2), 274–280.

Burney, D. A. (1993). Late Holocene environmental changes in arid southwestern Madagascar. *Quaternary Research, 40*(1), 98–106.

Burney, D. A., Hume, J. P., Randalana, R., Andrianaivoarivelo, R. A., Griffiths, O., Middleton, G. J., Rasolondrainy, T., Ramilisonina, & Radimilahy, C. (2020). Rock art from Andriamamelo Cave in the Beanka Protected Area of western Madagascar. *The Journal of Island and Coastal Archaeology,* 1–24.

Burney, D. A., & Ramilisonina. (1998). The *Kilopilopitsofy, Kidoky,* and *Bokyboky*: Accounts of strange animals from Belo-sur-mer, Madagascar, and the megafaunal 'extinction window'. *American Anthropologist, 100*(4), 957–966.

Burney, D. A., Robinson, G. S., & Burney, L. P. (2003). *Sporormiella* and the late Holocene extinctions in Madagascar. *Proceedings of the National Academy of Sciences, 100*(19), 10800–10805.

Burns, S. J., Godfrey, L. R., Faina, P., McGee, D., Hardt, B., Ranivoharimanana, L., & Randrianasy, J. (2016). Rapid human-induced landscape transformation in Madagascar at the end of the first millennium of the Common Era. *Quaternary Science Reviews, 134*, 92–99.

Caccone, A., Amato, G., Gratry, O. C., Behler, J., & Powell, J. R. (1999). A molecular phylogeny of four endangered Madagascar tortoises based on mtDNA sequences. *Molecular Phylogenetics and Evolution, 12*(1), 1–9.

Campbell, G. (2005). *An Economic History of Imperial Madagascar, 1750–1895: The Rise and Fall of an Island Empire* (Vol. 106). New York: Cambridge University Press.

Campbell, G. (2013). Forest Depletion in Imperial Madagascar, c. 1790–1861. In S. Evers, G. Campbell, & M. Lambek (Eds.), *Contest for Land in Madagascar: Environment, Ancestors, and Development* (Vol. 31, pp. 63–95). Leiden, the Netherlands: Brill.

Campbell, G. (2016). Africa and the Early Indian Ocean World Exchange System in the Context of Human–Environment Interaction. In G. Campbell (Ed.), *Early Exchange between Africa and the Wider Indian Ocean World* (pp. 1–24). Cham: Springer International Publishing.

Carroll, S. B. (2016). *The Serengeti Rules: The Quest to Discover How Life Works and Why it Matters*. Princeton, NJ: Princeton University Press.

Casson, L. (1989). *The Periplus Maris Erythraei: Text with Introduction, Translation, and Commentary*. Princeton: Princeton University Press.

Clarke, S. J., Miller, G. H., Fogel, M. L., Chivas, A. R., & Murray-Wallace, C. V. (2006). The amino acid and stable isotope biogeochemistry of elephant bird (*Aepyornis*) eggshells from southern Madagascar. *Quaternary Science Reviews, 25*(17), 2343–2356.

Cleese, J., & Johnstone, I. (Writers, from an idea by Jones, T. & Palin, M.). (1997). *Fierce Creatures* [Film]. In R. Young & F. Schepisi (Directors). Jersey Films.

Clements, F. E. (1916). Plant succession: An analysis of the development of vegetation. *Carnegie Institute of Washington*, Publication 242.

Codron, D., Codron, J., Sponheimer, M., Bernasconi, S. M., & Clauss, M. (2011). When animals are not quite what they eat: Diet digestibility influences 13C-incorporation rates and apparent discrimination in a mixed-feeding herbivore. *Canadian Journal of Zoology, 89*(6), 453–465.

Coe, M., Bourn, D., & Swingland, I. (1979). The biomass, production and carrying capacity of giant tortoises on Aldabra. *Philosophical Transactions of the Royal Society of London B: Biological Sciences, 286*(1011), 163–176.

Corson, C. A. (2016). *Corridors of Power: The Politics of Environmental Aid to Madagascar*. New Haven: Yale University Press.

Cox, R., Bierman, P., Jungers, M. C., & Rakotondrazafy, A. F. M. (2009). Erosion rates and sediment sources in Madagascar inferred from 10Be analysis of lavaka, slope, and river sediment. *The Journal of Geology, 117*(4), 363–376.

Cox, R., Zentner, D. B., Rakotondrazafy, A. F. M., & Rasoazanamparany, C. F. (2010). Shakedown in Madagascar: Occurrence of lavakas (erosional gullies) associated with seismic activity. *Geology, 38*(2), 179–182.

Crottini, A., Madsen, O., Poux, C., Strauss, A., Vieites, D. R., & Vences, M. (2012). Vertebrate time-tree elucidates the biogeographic pattern of a major biotic change around the K–T boundary in Madagascar. *Proceedings of the National Academy of Sciences, 109*(14), 5358–5363.

Crowley, B. E. (2010). A refined chronology of prehistoric Madagascar and the demise of the megafauna. *Quaternary Science Reviews, 29*(19), 2591–2603.

Crowley, B., Godfrey, L., Hansford, J., & Samonds, K. (2021). Seeing the forest for the trees – and the grasses: Revisiting the evidence for grazer-maintained grasslands in Madagascar's Central Highlands. *Proceedings of the Royal Society B, 288*, 20201785. doi: https://doi.org/10.1098/rspb.2020.1785

Crowley, B. E., & Samonds, K. E. (2013). Stable carbon isotope values confirm a recent increase in grasslands in northwestern Madagascar. *Holocene, 23*(7), 1066–1073. doi: 10.1177/0959683613484675

Crowther, A., Lucas, L., Helm, R., Horton, M., Shipton, C., Wright, H. T., Walshaw, S., Pawlowicz, M., Radimilahy, C., Douka, K., Picornell-Gelabert, L., Fuller, D. Q., & Boivin, N. L. (2016). Ancient crops provide first archaeological signature of the westward Austronesian expansion. *Proceedings of the National Academy of Sciences, 113*(24), 6635–6640. doi: 10.1073/pnas.1522714113

Dahl, O. C. (1951). *Malgache et Maanjan: Une Comparaison Linguistique* (Vol. 3). Oslo: A. Gimne.

Davis, D. S., Andriankaja, V., Carnat, T. L., Chrisostome, Z. M., Colombe, C., Fenomanana, F., Hubertine, L., Justome, R., Lahiniriko, F., Léonce, H., Manahira, G., Pierre, B.V., Razafimagnefa, R., Soafiavy, P., Victorian, F., Voahirana, V., Manjakahery, B., & Douglass, K. (2020). Satellite-based remote sensing rapidly reveals extensive record of Holocene coastal settlement on Madagascar. *Journal of Archaeological Science, 115*, 105097.

de Queiroz, A. (2014). *The Monkey's Voyage: How Improbable Journeys Shaped the History of Life*. New York: Basic Books.

de Wit, M. J. (2003). Madagascar: Heads it's a continent, tails it's an island. *Annual Review of Earth and Planetary Sciences, 31*(1), 213–248. doi: 10.1146/annurev.earth.31.100901.141337

de Wit, M. J., & Anderson, J. M. (2003). Gondwana alive corridors: Extending Gondwana research to incorporate stemming the sixth extinction. *Gondwana Research, 6*(3), 369–408. doi: 10.1016/s1342-937x(05)70994-1

Delumeau, J. (1995). *Une Histoire du Paradis, Tome 2: Mille Ans de Bonheur*. Paris: Fayard.

Deschamps, H. (1972). *Histoire de Madagascar*. Paris: Berger-Levrault.

Dewar, R. E. (1984). Extinctions in Madagascar: The loss of the Subfossil Fauna. In P. Martin & R. Klein (Eds.), *Quaternary Extinctions: A Prehistoric Revolution* (pp. 574–593). Tucson: University of Arizona Press.

Dewar, R. E. (1997). Were People Responsible for the Extinction of Madagascar's Subfossils, and How Will We Ever Know. In S. Goodman & B. Patterson (Eds.), *Natural Change and Human Impact in Madagascar* (pp. 364–377). Washington, D.C.: Smithsonian Institution Press.

Dewar, R. E. (2014). Early Human Settlers and Their Impact on Madagascar's Landscapes. In I. Scales (Ed.), *Conservation and Environmental Management in Madagascar* (pp. 44–64). London: Routledge.

Dewar, R. E., Radimilahy, C., Wright, H. T., Jacobs, Z., Kelly, G. O., & Berna, F. (2013). Stone tools and foraging in northern Madagascar challenge Holocene extinction models. *Proceedings of the National Academy of Sciences, 110*(31), 12583–12588. doi: 10.1073/pnas.1306100110

Dewar, R. E., & Richard, A. F. (2007). Evolution in the hypervariable environment of Madagascar. *Proceedings of the National Academy of Sciences, 104*(34), 13723–13727. doi: 10.1073/pnas.0704346104

Dewar, R. E., & Wallis, J. R. (1999). Geographical patterning of interannual rainfall variability in the tropics and near tropics: An L-moments approach. *Journal of Climate, 12*(12), 3457–3466. doi: 10.1175/1520-0442(1999)012<3457:Gpoirv>2.0.Co;2

Douglass, K. (2016). *An Archaeological Investigation of Settlement and Resource Exploitation Patterns in the Velondriake Marine Protected Area, Southwest Madagascar, ca. 900 BC to AD 1900.* (Ph.D. Dissertation), Yale University, New Haven.

Douglass, K., Hixon, S., Wright, H. T., Godfrey, L. R., Crowley, B. E., Manjakahery, B., Rasolondrainy, T., Crossland, Z., & Radimilahy, C. (2019). A critical review of radiocarbon dates clarifies the human settlement of Madagascar. *Quaternary Science Reviews, 221*, 105878.

Douglass, K., & Zinke, J. (2015). Forging ahead by land and by sea: Archaeology and paleoclimate reconstruction in Madagascar. *African Archaeological Review, 32*(2), 267–299.

Dove, M. R. (2017). Foreword. *Vanilla Landscapes* (pp. ix–x). New York: The New York Botanical Garden.

Eisenberg, J. F., & Gould, E. (1970). *The Tenrecs: A Study in Mammalian Behavior and Evolution*. Washington, D.C.: Smithsonian Institution Press.

Ekblom, A., & Gillson, L. (2010). Fire history and fire ecology of Northern Kruger (KNP) and Limpopo National Park (PNL), southern Africa. *The Holocene, 20*(7), 1063–1077.

Ekblom, A., Lane, P., Radimilahy, C., Rakotoarisoa, J-A., Sinclair, P., & Virah-Sawmy, M. (2016). Migration and Interaction between Madagascar and Eastern Africa, 500 BCE–1000 CE: An Archaeological Perspective. In G. Campbell (Ed.), *Early Exchange Between Africa and the Wider Indian Ocean World* (pp. 195–230). Cham: Springer International Publishing.

Eklund, J., Blanchet, F. G., Nyman, J., Rocha, R., Virtanen, T., & Cabeza, M. (2016). Contrasting spatial and temporal trends of protected area effectiveness in mitigating deforestation in Madagascar. *Biological Conservation, 203*, 290–297.

Ellis, S. (1985). *The Rising of the Red Shawls: A Revolt in Madagascar, 1895–1899*. Cambridge: Cambridge University Press.

Ellis, S. (2007). Tom and Toakafo: The Betsimisaraka kingdom and state formation in Madagascar, 1715–1750. *The Journal of African History, 48*(3), 439–455.

Elmqvist, T., Pyykonen, M., Tengo, M., Rakotondrasoa, F., Rabakonandrianina, E., & Radimilahy, C. (2007). Patterns of loss and regeneration of tropical dry forest in Madagascar: The social institutional context. *PLoS One, 2*(5), e402. doi: 10.1371/journal. pone.0000402

Eltringham, S. K. (1999). *The Hippos: Natural History and Conservation*. Cambridge: Cambridge University Press.

Erickson, C. J., Nowicki, S., Dollar, L., & Goehring, N. (1998). Percussive foraging: Stimuli for prey location by aye-ayes (*Daubentonia madagascariensis*). *International Journal of Primatology, 19*(1), 111–122.

Evans, S. E., Jones, M. E., & Krause, D. W. (2008). A giant frog with South American affinities from the Late Cretaceous of Madagascar. *Proceedings of the National Academy of Sciences, 105*(8), 2951–2956.

Evers, S., Campbell, G., & Lambek, M. (Eds.). (2013). *Contest for Land in Madagascar: Environment, Ancestors and Development.* Leiden, the Netherlands: Brill.

Everson, K. M., Hildebrandt, K. B., Goodman, S. M., & Olson, L. E. (2018). Caught in the act: Incipient speciation across a latitudinal gradient in a semifossorial mammal from Madagascar, the mole tenrec *Oryzorictes hova* (Tenrecidae). *Molecular Phylogenetics and Evolution, 126,* 74–84.

Faina, P., Burns, S. J., Godfrey, L. R., Crowley, B. E., Scroxton, N., McGee, D., Sutherland, M. R., & Ranivoharimanana, L. (2021). Comparing the paleoclimates of northwestern and southwestern Madagascar during the late Holocene: Implications for the role of climate in megafaunal extinction. *Malagasy Nature, 15.*

Fairhead, J., Leach, M., & Scoones, I. (2012). Green grabbing: A new appropriation of nature? *Journal of Peasant Studies, 39*(2), 237–261.

Federman, S., Donoghue, M. J., Daly, D. C., & Eaton, D. A. (2018). Reconciling species diversity in a tropical plant clade (*Canarium*, Burseraceae). *PLoS One, 13*(6), e0198882.

Feeley-Harnik, G. (2001). Ravenala Madagascariensis Sonnerat: The historical ecology of a 'flagship species' in Madagascar. *Ethnohistory, 48*(1–2), 31–86.

Fisher, B. L., & Robertson, H. G. (2002). Comparison and origin of forest and grassland ant assemblages in the High Plateau of Madagascar (Hymenoptera: Formicidae). *Biotropica, 34*(1), 155–167.

Flacourt, É. de (1658). *Histoire de la Grande Isle Madagascar.* Paris: P L'Amy.

Flannery, T. (1994). *The Future Eaters: An Ecological History of the Australasian Lands and People.* Chatswood, New South Wales: Reed Books.

Flannery, T. (2017). Where Song Began: Australia's Birds and How They Changed the World. *New York Review of Books, 64*(4), 27.

Flynn, J. J., & Wyss, A. R. (2002). Madagascar's mesozoic secrets. *Scientific American, 286*(2), 54.

Fuller, D. Q., & Boivin, N. (2009). Crops, cattle and commensals across the Indian Ocean. Current and potential archaeobiological evidence. *Etudes Océan Indien*, (42–43), 13–46.

Ganzhorn, J. U., Arrigo-Nelson, S., Boinski, S., Bollen, A., Carrai, V., Derby, A., Donati. G., Koenig, A., Kowalewski, M., Lahann, P., Norscia, I., Polowinsky, S. Y., Schwitzer, C., Stevenson, P. R., Talebi, M., Tan, C., Vogel, E. R., & Wright, P. C. (2009). Possible fruit protein effects on primate communities in Madagascar and the neotropics. *PLoS One, 4*(12), e8253. doi: 10.1371/journal. pone.0008253

Gardner, C. J., Gabriel, F. U., John, F. A. S., & Davies, Z. G. (2016). Changing livelihoods and protected area management: A case study of charcoal production in south-west Madagascar. *Oryx, 50*(3), 495–505.

Gardner, C. J., Nicoll, M. E., Mbohoahy, T., Oleson, K. L. L., Ratsifandrihamanana, A. N., Ratsirarson, J., René de Roland, L-A., Virah-Sawmy, M., Zafindrasilivonona, B., & Davies, Z. G. (2013). Protected areas for conservation and poverty alleviation: Experiences from Madagascar. *Journal of Applied Ecology, 50*(6), 1289–1294. doi: 10.1111/1365-2664.12164

Gasse, F., & Van Campo, E. (1998). A 40,000-yr pollen and diatom record from Lake Tritrivakely, Madagascar, in the southern tropics. *Quaternary Research, 49*(3), 299–311.

Gasse, F., & Van Campo, E. (2001). Late Quaternary environmental changes from a pollen and diatom record in the southern tropics (Lake Tritrivakely, Madagascar). *Palaeogeography, Palaeoclimatology, Palaeoecology, 167*(3–4), 287–308.

Gautier, E. F. (1902). *Madagascar. Essai de Géographie Physique.* Paris: A. Challamel: Librairie Maritime et Coloniale.

Gerlach, J., Muir, C., & Richmond, M. D. (2006). The first substantiated case of trans-oceanic tortoise dispersal. *Journal of Natural History, 40*(41–43), 2403–2408.

Giacoma, C., Sorrentino, V., Rabarivola, C., & Gamba, M. (2010). Sex differences in the song of *Indri indri. International Journal of Primatology, 31*(4), 539–551.

Gibbs, J. P., Sterling, E. J., & Zabala, F. J. (2010). Giant tortoises as ecological engineers: A long-term quasi-experiment in the Galápagos Islands. *Biotropica, 42*(2), 208–214.

Gibson, C., & Hamilton, J. (1983). Feeding ecology and seasonal movements of giant tortoises on Aldabra atoll. *Oecologia, 56*(1), 84–92.

Glaw, F., Köhler, J., Hawlitschek, O., Ratsoavina, F. M., Rakotoarison, A., Scherz, M. D., & Vences, M. (2021). Extreme miniaturization of a new amniote vertebrate and insights into the evolution of genital size in chameleons. *Scientific Reports, 11*(1), 1–14.

Glaw, F., & Raselimanana, A. P. (2018). Systématique des Reptiles Terrestres Malgaches (ordres: Squamata, Testudines et Crocodylia) / Systematics of Terrestrial Malagasy Reptiles (orders Squamata, Testudines, and Crocodylia). In S. Goodman, M. Raherilalao, & S. Wohlhauser (Eds.), *The Terrestrial Protected Areas of Madagascar: Their History, Description, and Biota* (Vol. 1, pp. 289–328). Antananarivo, Madagascar: Association Vahatra.

Glaw, F., & Vences, M. (2007). *A Field Guide to the Amphibians and Reptiles of Madagascar* (3rd ed.). Köln, Germany: Vences & Glaw.

Godfrey, L. R., & Crowley, B. (2016). Madagascar's ephemeral palaeo-grazer guild: Who ate the ancient C4 grasses? *Proceedings of the Royal Society B, 283.*

Godfrey, L. R., Crowley, B. E., Muldoon, K. M., Kelley, E. A., King, S. J., Best, A. W., & Berthaume, M. A. (2015). What did *Hadropithecus* eat, and why should paleoanthropologists care? *American Journal of Primatology,* 1–15.

Godfrey, L. R., & Jungers, W. (2003). Subfossil Lemurs. In S. Goodman & J. Benstead (Eds.), *The Natural History of Madagascar* (pp. 1247–1252). Chicago: University of Chicago Press.

Godfrey, L. R., Jungers, W., Reed, K., Simons, E., & Chatrath, P. (1997). Subfossil Lemurs: Inferences About Past and Present Primate Communities in Madagascar. In S. Goodman & B. Patterson (Eds.), *Natural Change and Human Impact in Madagascar.* Washington, D.C.: Smithsonian Institution Press.

Godfrey, L. R., & Rasoazanabary, E. (2012). Demise of the Bet Hedgers: A Case Study of Human Impacts on Past and Present Lemurs of Madagascar. In G. M. Sodikoff (Ed.), *The Anthropology of Extinction* (pp. 165–199). Bloomington, IN: Indiana University Press.

Godfrey, L. R., Scroxton, N., Crowley, B. E., Burns, S. J., Sutherland, M. R., Pérez, V. R., Faina, P., McGee, D., & Ranivoharimanana, L. (2019). A new interpretation of Madagascar's megafaunal decline: The 'Subsistence Shift Hypothesis'. *Journal of Human Evolution, 130*, 126–140.

Goodman, S. M. (2011). *Les Chauves-Souris de Madagascar.* Antananarivo, Madagascar: Association Vahatra.

Goodman, S. M. (2012). *Les Carnivora de Madagascar.* Antananarivo, Madagascar: Association Vahatra.

Goodman, S. M., & Benstead, J. P., (Eds.). (2003). *The Natural History of Madagascar.* Chicago: University of Chicago Press.

Goodman, S. M., & Jungers, W. L. (2014). *Extinct Madagascar: Picturing the Island's Past.* Chicago: University of Chicago Press.

Goodman, S. M., & Raherilalao, M. J. (2018). Systématique des oiseaux malgaches / Systematics of Malagasy birds. In S. Goodman, M. Raherilalao, & S. Wohlhauser (Eds.), *The Terrestrial Protected Areas of Madagascar: Their History, Description, and Biota* (Vol. 1, pp. 329–362). Antananarivo, Madagascar: Association Vahatra.

Goodman, S. M., Raherilalao, M. J., & Muldoon, K. (2013). Bird fossils from Ankilitelo Cave: Inference about Holocene environmental changes in Southwestern Madagascar. *Zootaxa, 3750*(5), 534–548.

Goodman, S., Raherilalao, M., & Wohlhauser, S. (Eds.). (2018). *The Terrestrial Protected Areas of Madagascar: Their History, Description, and Biota.* Antananarivo, Madagascar: Association Vahatra.

Goodman, S., & Ramasindrazana, B. (2018). Systématique des Chauves-Souris Malgaches (Ordre de Chiroptera) / Systematics of Malagasy Bats (Order Chiroptera). In S. Goodman, M. Raherilalao, & S. Wohlhauser (Eds.), *The Terrestrial Protected Areas of Madagascar:*

Their History, Description, and Biota (Vol. 1, pp. 383–394). Antananarivo, Madagascar: Association Vahatra.

Goodman, S. M., & Soarimalala, V. (2018). Systématique des Rongeurs Endémiques Malgaches (Famille des Nesomyidae : Sous-famille des Nesomyinae) / Systematics of Endemic Malagasy Rodents (Family Nesomyidae: Subfamily Nesomyinae). In S. Goodman, M. Raherilalao, & S. Wohlhauser (Eds.), *The Terrestrial Protected Areas of Madagascar: Their History, Description, and Biota* (Vol. 1, pp. 373–382). Antananarivo, Madagascar: Association Vahatra.

Goodman, S. M., Soarimalala, V., & Olson, L. (2018). Systématique des Tenrecs Endémiques Malgaches (Famille des Tenrecidae) / Systematics of Endemic Malagasy Tenrecs (Family Tenrecidae). In S. Goodman, M. Raherilalao, & S. Wohlhauser (Eds.), *The Terrestrial Protected Areas of Madagascar: Their History, Description, and Biota* (Vol. 1, pp. 363–372). Antananarivo, Madagascar: Association Vahatra.

Graeber, D. (2007). *Lost people: Magic and the Legacy of Slavery in Madagascar*. Bloomington: Indiana University Press.

Grandidier, A. (1898). Le boisement de l'Imerina. *Bulletin du Comité de Madagascar, 4*(2), 83–87.

Grandidier, A., & Grandidier, G. (1904). *Madagascar: Collection des Ouvrages Anciens Concernant* (Vol. 2). Paris: Comité de Madagascar.

Gray, J. (2002). *Straw Dogs: Thoughts on Humans and Other Animals*. New York: Farrar, Straus & Giroux.

Griffiths, C. J., Zuël, N., Jones, C. G., Ahamud, Z., & Harris, S. (2013). Assessing the potential to restore historic grazing ecosystems with tortoise ecological replacements. *Conservation Biology, 27*(4), 690–700.

Grove, C. A., Zinke, J., Peeters, F., Park, W., Scheufen, T., Kasper, S., Randriamanantsoa, B., McCulloch, M. T., & Brummer, G-J. A. (2013). Madagascar corals reveal a multidecadal signature of rainfall and river runoff since 1708. *Climate of the Past, 9*(2), 641–656.

Grove, R. H. (1995). *Green Imperialism: Colonial Expansion, Tropical Island Edens and the Origins of Environmentalism, 1600–1860*. Cambridge, U. K.: Cambridge University Press.

Gunnell, G. F., Boyer, D. M., Friscia, A. R., Heritage, S., Manthi, F. K.,

Miller, E. R., Sallam, H. M., Simmons, N. B., Stevens, N. J., & Seiffert, E. R. (2018). Fossil lemurs from Egypt and Kenya suggest an African origin for Madagascar's aye-aye. *Nature communications, 9*(1), 3193.

Haeckel, E. (1876). *The History of Creation: On the Development of the Earth and Its Inhabitants by the Action of Natural Causes; a Popular Exposition of the Doctrine of Evolution in General, and of that of Darwin, Goethe, and Lamarck In Particular* (E. Lankester, Trans.). New York: D. Appleton and Co.

Hankel, O. (1993). Early Triassic plant microfossils from Sakamena sediments of the Majunga Basin, Madagascar. *Review of Palaeobotany and Palynology, 77*(3–4), 213–233.

Hansford, J. P., & Turvey, S. T. (2018). Unexpected diversity within the extinct elephant birds (Aves: Aepyornithidae) and a new identity for the world's largest bird. *Open Science, 5*(9), 181295.

Hansford, J., & Turvey, S. T. (in press). Reconstructing the dietary ecology of aepyornithids using stable isotope analysis. *Biology Letters.*

Hansford, J., Wright, P. C., Rasoamiaramanana, A., Pérez, V. R., Godfrey, L. R., Errickson, D., Thompson, T., & Turvey, S. T. (2018). Early Holocene human presence in Madagascar evidenced by exploitation of avian megafauna. *Science Advances, 4*(9).

Harper, G. J., Steininger, M. K., Tucker, C. J., Juhn, D., & Hawkins, F. (2007). Fifty years of deforestation and forest fragmentation in Madagascar. *Environmental Conservation, 34*(4), 325–333.

Harris, A. (2007). 'To live with the sea': Development of the Velondriake community-managed protected area network, Southwest Madagascar. *Madagascar Conservation & Development, 2*(1), 43–49.

Hawkins, A., & Goodman, S. (2003). Introduction to the Birds. In S. Goodman & J. Benstead (Eds.), *The Natural History of Madagascar* (pp. 1019–1044). Chicago: University of Chicago Press.

Heckman, K. L., Rasoazanabary, E., Machlin, E., Godfrey, L. R., & Yoder, A. D. (2006). Incongruence between genetic and morphological diversity in *Microcebus griseorufus* of Beza Mahafaly. *BMC Evolutionary Biology, 6*(1), 98.

Hekkala, E., Gatesy, J., Narechania, A., Meredith, R., Russello, M.,

Aardema, M. L., Jensen, E., Montanari, S., Brochu, C., Norell, M., & Amato, G. (2021). Paleogenomics illuminates the evolutionary history of the extinct Holocene 'horned' crocodile of Madagascar, *Voay robustus*. *Communications Biology, 4*(1), 505. doi: 10.1038/s42003-021-02017-0

Hixon, S., Curtis, J., Brenner, M., Douglass, K., Domic, A., Culleton, B., Ivory, S., & Kennett, D. (2021). Drought coincided with, but does not explain, late Holocene megafauna extinctions in SW Madagascar. *Climate, 9*(9), 138.

Hobman, B., & Burley, I. (1989). *Sarimanok: De Bali à Madagascar: Dans le Village des Marins de la Préhistoire*. Paris: B. Grasset.

Hoffmann, S., Hu, Y., & Krause, D. W. (2020). Postcranial morphology of *Adalatherium hui* (Mammalia, Gondwanatheria) from the late Cretaceous of Madagascar. *Journal of Vertebrate Paleontology, 40*(sup1), 133–212.

Horton, M., Boivin, N., Crowther, A., Gaskell, B., Radimilahy, C., & Wright, H. (2017). East Africa as a Source for Fatimid Rock Crystal. Workshops from Kenya to Madagascar. In A. Hilgner, S. Greiff, & D. Quast (Eds.), *Gemstones in the First Millennium AD: Mines, Trade, Workshops and Symbolism* (pp. 103–118). Mainz: Verlag des Romisch-Germanisches Zentralmuseums.

Humbert, H. (1958). Henri Perrier de la Bathie (1873–1958). *Journal d'Agriculture Tropicale et de Botanique Appliquée*, 863–867.

Jablonski, D. (1995). Extinction in the Fossil Record. In J. Lawton & R. May (Eds.), *Extinction Rates* (pp. 25–44). Oxford: Oxford University Press.

Jacobs, B. F., Kingston, J. D., & Jacobs, L. L. (1999). The origin of grass-dominated ecosystems. *Annals of the Missouri Botanical Garden, 86*(2), 590–643. doi: Doi 10.2307/2666186

Jarosz, L. (1993). Defining and explaining tropical deforestation: Shifting cultivation and population growth in colonial Madagascar (1896–1940). *Economic Geography, 69*(4), 366–379.

Jolly, A. (1980). *A World Like our Own: Man and Nature in Madagascar*. New Haven: Yale University Press.

Jolly, A. (1987). Madagascar: A world apart. *National Geographic, 171*(2), 148–183.

Jolly, A. (2004). *Lords and Lemurs: Mad Scientists, Kings with Spears, and the Survival of Diversity in Madagascar.* Boston: Houghton Mifflin Co.

Jolly, A. (2015). *Thank You, Madagascar: The Conservation Diaries of Alison Jolly.* London: Zed Books Ltd.

Jones, J., Andriamarovololona, M. M., & Hockley, N. (2008). The importance of taboos and social norms to conservation in Madagascar. *Conservation Biology, 22*(4), 976–986.

Jones, J., Rakotonarivo, O., & Razafimanahaka, J. (in press). Forest Conservation in Madagascar: Past, Present, and Future. In S. M. Goodman (Ed.), *The New Natural History of Madagascar* (2nd ed.). Princeton: Princeton University Press.

Jones, J., Ratsimbazafy, J., Ratsifandrihamanana, A. N., Watson, J. E., Andrianandrasana, H. T., Cabeza, M., Cinner, J. E., Goodman, S. M., Hawkins, F., Mittermeier, R. A., Rabearisoa, A. L., Rakotonarive, O. S., Razafimanahaka, J. H., Razafimpahanana, A. R., Wilmé, L., & Wright, P. C. (2019). Last chance for Madagascar's biodiversity. *Nature Sustainability, 2*(5), 350–352.

Joseph, G. S., & Seymour, C. L. (2020). Madagascan highlands: Originally woodland and forest containing endemic grasses, not grazing-adapted grassland. *Proceedings of the Royal Society B, 287*(1937), 20201956.

Kammerer, C. F., Nesbitt, S. J., Flynn, J. J., Ranivoharimanana, L., & Wyss, A. R. (2020). A tiny ornithodiran archosaur from the Triassic of Madagascar and the role of miniaturization in dinosaur and pterosaur ancestry. *Proceedings of the National Academy of Sciences, 117*(30), 17932–17936.

Kappeler, P. M., & Schäffler, L. (2008). The lemur syndrome unresolved: Extreme male reproductive skew in sifakas (*Propithecus verreauxi*), a sexually monomorphic primate with female dominance. *Behavioral Ecology and Sociobiology, 62*(6), 1007–1015.

Karsten, K. B., Andriamandimbiarisoa, L. N., Fox, S. F., & Raxworthy, C. J. (2008). A unique life history among tetrapods: An annual chameleon living mostly as an egg. *Proceedings of the National Academy of Sciences, 105*(26), 8980–8984.

Kaufmann, J. C. (2014). Contrasting Visions of Nature and Landscapes. In I. Scales (Ed.), *Conservation and Environmental Management in Madagascar* (pp. 320–341). London: Routledge.

Keller, E. (2015). *Beyond the Lens of Conservation: Malagasy and Swiss Imaginations of One Another* (Vol. 20). New York: Berghahn Books.

Koechlin, J., Guillaumet, J-L., & Morat, P. (1974). *Flore et Végétation de Madagascar*. Vaduz, Liechtenstein: J. Cramer.

Krause, D. W. (2001). Fossil molar from a Madagascan marsupial – The discovery of a tiny tooth from the Late Cretaceous period has sizeable implications. *Nature, 412*(6846), 497–498. doi: Doi 10.1038/35087649

Krause, D. W. (2003). Late Cretaceous Vertebrates of Madagascar: A Window into Gondwanan Biogeography at the End of the Age of Dinosaurs. In S. Goodman & J. Benstead (Eds.), *The Natural History of Madagascar* (pp. 40–47). Chicago: University of Chicago Press.

Krause, D. W., Hartman, J., & Wells, N. (1997). Late Cretaceous Vertebrates from Madagascar: Implications for Biotic Change in Deep Time. In S. Goodman & B. Patterson (Eds.), *Natural Change and Human Impact in Madagascar* (pp. 3–43). Washington, D.C.: Smithsonian Institution Press.

Krause, D. W., Hoffmann, S., Hu, Y., Wible, J., Rougier, G., Kirk, E., Groenke, J. R., Rogers, R. R., Rossie, J. B., Schultz, J. A., Evans, A. R., von Koenigswald, W., & Rahantarisoa, L. J. (2020). Skeleton of a Cretaceous mammal from Madagascar reflects long-term insularity. *Nature, 581*(7809), 421–427.

Krause, D. W., Hoffmann, S., Wible, J. R., Kirk, E. C., Schultz, J. A., von Koenigswald, W., Groenke, J. R., Rossie, J. B., O'Connor, P. M., Seiffert, E. R., Dumont, E. R., Holloway, W. L., Rogers, R. R., Rahantarisoa, L. J., Kemp, A. D., & Andriamialison, H. (2014). First cranial remains of a Gondwanatherian mammal reveal remarkable mosaicism. *Nature, 515*(7528), 512–517.

Kress, W. J., Schatz, G. E., Andrianifahanana, M., & Morland, H. S.

(1994). Pollination of *Ravenala madagascariensis* (Strelitziaceae) by lemurs in Madagascar: Evidence for an archaic coevolutionary system? *American Journal of Botany, 81*(5), 542–551.

Kull, C. A. (2000). Deforestation, erosion, and fire: Degradation myths in the environmental history of Madagascar. *Environment and History, 6*(4), 423–450.

Kull, C. A. (2004). *Isle of Fire: The Political Ecology of Landscape Burning in Madagascar.* Chicago: University of Chicago Press.

Kull, C. A. (2014). The Roots, Persistence, and Character of Madagascar's Conservation Boom. In I. Scales (Ed.), *Conservation and Environmental Management in Madagascar* (pp. 146–171). London: Routledge.

Kull, C. A., Tassin, J., & Carrière, S. M. (2014). Approaching invasive species in Madagascar. *Madagascar Conservation & Development, 9*(2), 60–70.

Kusuma, P., Brucato, N., Cox, M. P., Pierron, D., Razafindrazaka, H., Adelaar, A., Sudoyo, H., Letellier, T., & Ricaut, F-X. (2016). Contrasting linguistic and genetic origins of the Asian source populations of Malagasy. *Scientific Reports, 6*, 26066.

Kusuma, P., Cox, M. P., Pierron, D., Razafindrazaka, H., Brucato, N., Tonasso, L., Suryadi, H. L., Letellier, T., Sudoyo, H., & Ricaut, F-X. (2015). Mitochondrial DNA and the Y chromosome suggest the settlement of Madagascar by Indonesian sea nomad populations. *BMC Genomics, 16*(1), 191.

Lamberton, C. (1934). *Contribution à la Connaissance de la Faune Subfossile de Madagascar.* Tananarive, Madagascar: Imprimerie Moderne de l'Emyrne, G. Pitot & cie.

Lamberton, C. (1934). *Contribution à la Connaissance de la Faune Subfossile de Madagascar.* Tananarive, Madagascar: Imprimerie Moderne de l'Emyrne, G. Pitot & cie.

Langrand, O. (1990). *Guide to the Birds of Madagascar.* New Haven: Yale University Press.

Lappin, A. K., Wilcox, S. C., Moriarty, D. J., Stoeppler, S. A., Evans, S. E., & Jones, M. E. (2017). Bite force in the horned frog (*Ceratophrys cranwelli*) with implications for extinct giant frogs. *Scientific Reports*, 7(1), 11963.

Larson, P. M. (1995). Multiple narratives, gendered voices: Remembering the past in Highland Central Madagascar. *The International Journal of African Historical Studies, 28*(2), 295–325.

Larson, P. M. (2000). *History and Memory in the Age of Enslavement: Becoming Merina in Highland Madagascar, 1770–1822*. Portsmouth, NH: Heinemann.

Larson, P. M. (2009). *Ocean of Letters: Language and Creolization in an Indian Ocean Diaspora*. Cambridge: Cambridge University Press.

Lawler, R. R. (2008). Testing for a historical population bottleneck in wild Verreaux's sifaka (*Propithecus verreauxi verreauxi*) using microsatellite data. *American Journal of Primatology, 70*(10), 990–994.

Lehmann, C. E., Anderson, T. M., Sankaran, M., Higgins, S. I., Archibald, S., Hoffmann, W. A., Hanan, N. P., Williams, R. J., Fensham, R. J., Felfili, J., Hutley, L. B., Ratnam, J., Jose, J. S., Montes, R., Franklin, D., Russell–Smith, J., Ryan, C. M., Durigan, G., Hiernaux, P., Haidar, R., Bowman, D. M. J. S., & Bond, W. J. (2014). Savanna vegetation-fire-climate relationships differ among continents. *Science, 343*(6170), 548–552.

Lévi-Strauss, C. (1979). *Myth and Meaning*. New York: Schocken Books.

Lovejoy, T. (2013). A tsunami of extinction. *Scientific American, 308*(1), 33–34.

Low, T. (2016). *Where Song Began: Australia's Birds and How They Changed the World*. New Haven: Yale University Press.

Lowry II, P., Phillipson, P., Andriamahefarivo, L., Schatz, G., Rajaonary, F., & Andriambololonera, S. (2018). Flore / Flora. In S. Goodman, M. Raherilalao, & S. Wohlhauser (Eds.), *The Terrestrial Protected Areas of Madagascar: Their History, Description, and Biota* (Vol. 1, pp. 243–256). Antananarivo, Madagascar: Association Vahatra.

Mack, J. (2007). The land viewed from the sea. *Azania: Journal of the British Institute in Eastern Africa, 42*(1), 1–14.

MacPhee, R. D. (1994). Morphology, adaptations, and relationships of *Plesiorycteropus*, and a diagnosis of a new order of eutherian mammals. *Bulletin of the American Museum of Natural History, 220,* 1–214.

Mahe, J., & Sourdat, M. (1972). Sur l'extinction des vertébrés subfossiles et l'aridification du climat dans le Sud-ouest de Madagascar. *Bulletin de la Société de Géologie de France, 14,* 295–309.

Manguin, P-Y. (2016). Austronesian Shipping in the Indian Ocean: From Outrigger Boats to Trading Ships. In G. Campbell (Ed.), *Early Exchange between Africa and the Wider Indian Ocean World* (pp. 51–76). Cham: Springer International Publishing.

Marden, L. (1967). Madagascar: Island at the end of the Earth. *National Geographic Magazine, 132*(4), 443–487.

Martin, R. D. (1972). Review lecture: Adaptive radiation and behaviour of the Malagasy lemurs. *Philosophical Transactions of the Royal Society of London, Series B, Biological Sciences, 264*(862), 295–352.

Masters, J. C., Génin, F., Zhang, Y., Pellen, R., Huck, T., Mazza, P. A., Rabineau, M., Doucouré, M., & Aslanian, D. (2021). Biogeographic mechanisms involved in the colonization of Madagascar by African vertebrates: Rifting, rafting and runways. *Journal of Biogeography, 48,* 492–510.

Matsumoto, K., & Burney, D. A. (1994). Late Holocene environmental changes at Lake Mitsinjo, northwestern Madagascar. *The Holocene, 4*(1), 16–24.

Mavume, A. F., Rydberg, L., Rouault, M., & Lutjeharms, J. R. (2009). Climatology and landfall of tropical cyclones in the south-west Indian Ocean. *Western Indian Ocean Journal of Marine Science, 8*(1).

McConnell, W., & Kull, C. (2014). Deforestation in Madagascar: Debates Over the Island's Forest Cover and the Challenges of Measuring Forest Change. In I. Scales (Ed.), *Conservation and Environmental Management in Madagascar* (pp. 67–104). London, UK: Earthscan from Routledge.

McConnell, W. J., & Sweeney, S. P. (2005). Challenges of forest governance in Madagascar. *The Geographical Journal, 171*(3), 223–238.

McConnell, W. J., Viña, A., Kull, C., & Batko, C. (2015). Forest

transition in Madagascar's highlands: Initial evidence and implications. *Land, 4*(4), 1155–1181.

Meador, L. R., Godfrey, L. R., Rakotondramavo, J. C.,
Ranivoharimanana, L., Zamora, A., Sutherland, M. R., & Irwin,
M. T. (2019). *Cryptoprocta spelea* (Carnivora: Eupleridae): What did
it eat and how do we know? *Journal of Mammalian Evolution, 26*(2),
237–251.

Merton, L., Bourn, D., & Hnatiuk, R. (1976). Giant tortoise and
vegetation interactions on Aldabra Atoll – Part 1: Inland. *Biological
Conservation, 9*(4), 293–304.

Middleton, K. (1999). Who killed 'Malagasy Cactus'? Science,
environment and colonialism in southern Madagascar (1924–
1930). *Journal of Southern African Studies, 25*(2), 215–248. doi:
10.1080/030570799108678

Milton, S. J., Dean, W. R. J., & Siegfried, W. R. (1994). Food selection
by ostrich in southern Africa. *The Journal of Wildlife Management,
58*(2), 234–248.

Mitchell, P. J. (1996). Prehistoric exchange and interaction in southeastern southern Africa: Marine shells and ostrich eggshell. *African
Archaeological Review, 13*(1), 35–76.

Mittermeier, R. A., Louis, E. E., Langrand, O., Schwitzer, C.,
Gauthier, C-A., Rylands, A. B., Rajaobelina, S., Ratsimbazafy, J.,
Rasoloarison, R., Hawkins, F., Roos, C., Richardson, M., &
Kappeler, P. M. (2014). *Lemuriens de Madagascar.* Conservation
International, Arlington, VA: Publications Scientifiques du Museum
National d'Histoire Naturelle, Paris.

Mittermeier, R. A., & Nash, S. D. (2010). *Lemurs of Madagascar* (3rd
ed.). Arlington, VA: Conservation International.

Moat, J., & Smith, P. P. (2007). *Atlas of the Vegetation of Madagascar.*
Royal Botanic Gardens, Kew: Kew Publishing.

Muldoon, K., de Blieux, D., Simons, E., & Chatrath, P. (2009). The
subfossil occurrence and paleoecological significance of small
mammals at Ankilitelo Cave, southwestern Madagascar. *Journal of
Mammalogy, 90*(5), 1111–1131.

Myers, N. (1988). Threatened biotas: 'Hot spots' in tropical forests. *The Environmentalist, 8*(3), 187–208.

Nicoll, M. E. (2003). *Tenrec Ecaudatus,* Tenrec, *Tandraka, Trandraka.* In S. Goodman & J. Benstead (Eds.), *The Natural History of Madagascar* (pp. 1283–1287). Chicago: University of Chicago Press.

Nobre, T., Eggleton, P., & Aanen, D. (2010). Vertical transmission as the key to the colonization of Madagascar by fungus-growing termites? *Proceedings of the Royal Society of London B: Biological Sciences, 277,* 359–365.

Noonan, B. P., & Chippindale, P. T. (2006). Vicariant origin of Malagasy reptiles supports Late Cretaceous Antarctic land bridge. *The American Naturalist, 168*(6), 730–741.

Norell, M. A. (1992). Taxic Origin and Temporal Diversity: The Effect of Phylogeny. In M. J. Novacek & Q. D. Wheeler (Eds.), *Extinction and Phylogeny* (pp. 89–118). New York: Columbia University Press.

Nowack, J., & Dausmann, K. H. (2015). Can heterothermy facilitate the colonization of new habitats? *Mammal Review, 45*(2), 117–127. doi: 10.1111/mam.12037

O'Connor, P. M., & Forster, C. A. (2010). A Late Cretaceous (Maastrichtian) avifauna from the Maevarano Formation, Madagascar. *Journal of Vertebrate Paleontology, 30*(4), 1178–1201.

Ohba, M., Samonds, K., LaFleur, M., Ali, J. R., & Godfrey, L. R. (2016). Madagascar's climate at the K/P boundary and its impact on the island's biotic suite. *Palaeogeography Palaeoclimatology Palaeoecology, 441,* 688–695. doi: 10.1016/j.palaeo.2015.10.028

Okullo, P., & Moe, S. R. (2012). Termite activity, not grazing, is the main determinant of spatial variation in savanna herbaceous vegetation. *Journal of Ecology, 100*(1), 232–241.

Osterhoudt, S. R. (2017). *Vanilla Landscapes: Meaning, Memory, and the Cultivation of Place in Madagascar.* Bronx, New York: New York Botanical Garden.

Osterhoudt, S. R. (2020). 'Nobody wants to kill': Economies of affect and violence in Madagascar's vanilla boom. *American Ethnologist, 47*(3), 249–263.

Osterhoudt, S. R. (2021). Bright spot ethnography: On the analytical potential of things that work. *The Arrow: A Journal of Wakeful Society, Culture and Politics, 8*(1), 33–47.

Owen, R. (1863). On the aye–aye (*Chiromys*, Cuvier; *Chiromys madagascariensis*, Desm.; *Sciurus madagascariensis*, Gmel., Sonnerat; *Lemur psilodactylus*, Schreber, Shaw). *Transactions of the Zoological Society of London, 5*(2), 33–101.

Palkovacs, E. P., Gerlach, J., & Caccone, A. (2002). The evolutionary origin of Indian Ocean tortoises (*Dipsochelys*). *Molecular Phylogenetics and Evolution, 24*(2), 216–227.

Parker Pearson, M. (2010). *Pastoralists, Warriors and Colonists: The Archaeology of Southern Madagascar.* Oxford: Archaeopress.

Perez, V. R., Godfrey, L. R., Nowak–Kemp, M., Burney, D. A., Ratsimbazafy, J., & Vasey, N. (2005). Evidence of early butchery of giant lemurs in Madagascar. *Journal of Human Evolution, 49*(6), 722–742.

Perrier de la Bâthie, H. (1921). *La Végétation Malgache* (Vol. 9). Marseille: Musée Colonial.

Perrier de la Bâthie, H. (1936). *Biogéographie des Plantes de Madagascar.* Paris: Société d'Éditions Géographiques, Maritimes et Coloniales.

Pierron, D., Heiske, M., Razafindrazaka, H., Rakoto, I., Rabetokotany, N., Ravololomanga, B., Rakotozafy, L. M-A., Rakotomalala, M. M., Razafiarivony, M., Rasoarifetra, B., Raharijesy, M. A., Razafindralambo, L., Ramilisonina, Fanony, F., Lejamble, S., Thomas, O., Abdallah, A. M., Rocher, C., Arachiche, A., Tonaso, L., Pereda-lotha, V., Schiavinatoa, S., Brucato, N., Ricaut, F-X., Kusuma, P., Sudoyo, H., Ni, S., Boland, A., Deleuze, J-F., Beaujard, P., Grange, P., Adelaar, S., Stoneking, M., Rakotoarisoa, J-A., Radimilahy, C., & Letellier, T. (2017). Genomic landscape of human diversity across Madagascar. *Proceedings of the National Academy of Sciences, 114*(32), E6498–E6506.

Pollini, J. (2009). Agroforestry and the search for alternatives to slash-and-burn cultivation: From technological optimism to a political economy of deforestation. *Agriculture, Ecosystems & Environment, 133*(1–2), 48–60.

Pollini, J., Hockley, N., Muttenzer, F. D., & Ramamonjisoa, B. S. (2014). The Transfer of Natural Resource Management Rights to Local Communities. In I. Scales (Ed.), *Conservation and Environmental Management in Madagascar* (pp. 172–192). London: Routledge.

Pollock, J. I. (1986). The song of the indris (*Indri indri*; Primates: Lemuroidea): Natural history, form, and function. *International Journal of Primatology, 7*(3), 225–264.

Poudyal, M., Jones, J. P., Rakotonarivo, O. S., Hockley, N., Gibbons, J. M., Mandimbiniaina, R., Rasoamanana, A., Andrianantenaina, N. S., & Ramamonjisoa, B. S. (2018). Who bears the cost of forest conservation? *PeerJ, 6*, e5106.

Powzyk, J., & Thalmann, U. (2003). Indri *Indri indri*. In S. Goodman & J. Benstead (Eds.), *The Natural History of Madagascar* (pp. 1342–1345). Chicago: University of Chicago Press.

Prosek, J. (2020). *James Prosek: Art, Artifact, Artifice.* New Haven: Yale University Press.

Quammen, D. (2008). *Natural Acts: A Sidelong View of Science & Nature.* New York: W.W. Norton & Company.

Quéméré, E., Amelot, X., Pierson, J., Crouau-Roy, B., & Chikhi, L. (2012). Genetic data suggest a natural prehuman origin of open habitats in northern Madagascar and question the deforestation narrative in this region. *Proceedings of the National Academy of Sciences, 109*(32), 13028–13033.

Rabesahala Horning, N. (2003). How Rules Affect Conservation Outcomes. In S. Goodman & J. Benstead (Eds.), *The Natural History of Madagascar* (pp. 146–153). Chicago: University of Chicago Press.

Radimilahy, C. (1998). *Mahilaka: An Archaeological Investigation of an Early Town in Northwestern Madagascar.* (PhD), Acta Universitatis Upsaliensis, Uppsala.

Rafferty, A. R., & Reina, R. D. (2012). Arrested embryonic development: A review of strategies to delay hatching in egg-laying reptiles. *Proceedings of the Royal Society of London B: Biological Sciences, 279*(1737), 2299–2308.

Rakotoarisoa, J-A. (2002). Madagascar: Background Notes. In

C. Kreamer & S. Fee (Eds.), *Objects as Envoys: Cloth, Imagery, and Diplomacy in Madagascar* (pp. 25–32). Washington, DC: Smithsonian Institution, National Museum of African Art.

Rakotondrafara, M., Randriamarolaza, L., Rasolonjatovo, H., Rakotomalala, C., & Razanakiniana, F. (2018). Historical Evolution of Climatic Aspects on Madagascar Protected Areas. In S. Goodman, M. Raherilalao, & S. Wohlhauser (Eds.), *The Terrestrial Protected Areas of Madagascar: Their History, Description, and Biota* (Vol. 1, pp. 199–206). Antananarivo, Madagascar: Association Vahatra.

Rakotovao, M., Lignereux, Y., Orliac, M.J., Duranthon, F., & Antoine, P-O. (2014). Hippopotamus lemerlei Grandidier, 1868 et Hippopotamus madagascariensis Guldberg, 1883 (Mammalia, Hippopotamidae): Anatomie crânio-dentaire et révision systématique. *Geodiversitas, 36*(1), 117–161.

Ranaivonasy, J., Ratsirarson, J., Rasamimanana, N., & Ramahatratra, E. (2016). Dynamique de la couverture forestière dans la Réserve Spéciale de Bezà Mahafaly et ses environs. *Malagasy Nature, 10*, 15–24.

Ranaivonasy, J., Ratsirarson, J., & Richard, A. (2016). (Eds.). Suivi écologique et socio-économique dans la Réserve Spéciale de Bezà Mahafaly (Sud-ouest de Madagascar). *Malagasy Nature, 10.*

Randriamalala, H., & Liu, Z. (2010). Rosewood of Madagascar: Between democracy and conservation. *Madagascar Conservation & Development, 5*(1), 11–22.

Randrianandrianina, F. H., Racey, P. A., & Jenkins, R. K. (2010). Hunting and consumption of mammals and birds by people in urban areas of western Madagascar. *Oryx, 44*(3), 411–415.

Raombana. (1980). *Histoires* (S. Ayache, Trans. & Ed.). Fianarantsoa: Librairie Ambozontany.

Raper, D., & Bush, M. (2009). A test of *Sporormiella* representation as a predictor of megaherbivore presence and abundance. *Quaternary Research, 71*(3), 490–496.

Rasamuel, D. (1984). Alimentation et techniques anciennes dans le sud Malgache à travers une fosse à ordures du XIe siècle. *Etudes Océan Indien, 4*, 81–109.

Raselimanana, A. (2003). Trade in Reptiles and Amphibians. In S. Goodman & B. JP (Eds.), *The Natural History of Madagascar* (pp. 1564–1568). Chicago: University of Chicago Press.

Raselimanana, A., & Rakotomalala, D. (2003). *Chamaeleonidae,* Chameleons. In S. Goodman & J. Benstead (Eds.), *The Natural History of Madagascar* (pp. 960–969). Chicago: University of Chicago Press.

Rasoazanabary, E. (2011). *The Human Factor in Mouse Lemur Conservation: Local Resource Utilization and Habitat Disturbance at Bezà Mahafaly Special Reserve, SW Madagascar.* (Ph.D.), University of Massachusetts, Amherst.

Rasolondrainy, T. V. R. (2012). Discovery of rock paintings and Libyco-Berber inscription from the Upper Onilahy, Isalo Region, southwestern Madagascar. *Studies in the African Past, 10,* 173–195.

Ratovonamana, Y., Rajeriarison, C., Roger, E., Kiefer, I., & Ganzhorn, J. U. (2013). Impact of livestock grazing on forest structure, plant species composition and biomass in southwestern Madagascar. *Scripta Botanica Belgica, 50,* 82–92.

Raxworthy, C. (1991). Field observations on some dwarf chameleons (*Brookesia spp.*) from rainforest areas of Madagascar, with description of a new species. *Journal of Zoology, 224*(1), 11–25.

Raxworthy, C., Forstner, M., & Nussbaum, R. (2002). Chameleon radiation by oceanic dispersal. *Nature, 415*(6873), 784–787.

Razafimanahaka, J. H., Jenkins, R. K., Andriafidison, D., Randrianandrianina, F., Rakotomboavonjy, V., Keane, A., & Jones, J. P. (2012). Novel approach for quantifying illegal bushmeat consumption reveals high consumption of protected species in Madagascar. *Oryx, 46*(4), 584–592.

Razafindratsima, O. H., Jones, T. A., & Dunham, A. E. (2014). Patterns of movement and seed dispersal by three lemur species. *American Journal of Primatology, 76*(1), 84–96.

Reuter, K. E., Randell, H., Wills, A. R., Janvier, T. E., Belalahy, T. R., & Sewall, B. J. (2016). Capture, movement, trade, and consumption of mammals in Madagascar. *PLoS One, 11*(2), e0150305.

Richard, A. F., & Dewar, R. E. (1991). Lemur ecology. *Annual Review of Ecology and Systematics, 22*, 145–175. doi: DOI 10.1146/annurev. ecolsys.22.1.145

Richard, A. F., & Dewar, R. E. (2001). Politics, Negotiation, and Conservation: A View from Madagascar. In W. Weber, L. White, A. Vedder, & L. Naughton-Treves (Eds.), *African Rain Forest Ecology and Conservation: An Interdisciplinary Perspective* (pp. 535–544). New Haven: Yale University Press.

Richard, A. F., Goldstein, S. J., & Dewar, R. E. (1989). Weed macaques – The evolutionary implications of macaque feeding ecology. *International Journal of Primatology, 10*(6), 569–597. doi: Doi 10.1007/ Bf02739365

Richard, A. F., & Nicoll, M. (1987). Female social-dominance and basal-metabolism in a Malagasy primate, *Propithecus verreauxi*. *American Journal of Primatology, 12*(3), 309–314. doi: DOI 10.1002/ ajp.1350120308

Richard, A. F., & O'Connor, S. (1997). Degradation, Transformation, and Conservation: The Past, Present, and Possible Future of Madagascar's Environment. In S. Goodman & B. Patterson (Eds.), *Natural Change and Human Impact in Madagascar* (pp. 406–418). Washington, DC: Smithsonian Institution Press.

Richard, A. F., & Ratsirarson, J. (2013). Partnership in practice: Making conservation work at Bezà Mahafaly, southwest Madagascar. *Madagascar Conservation & Development, 8*(1), 12–20.

Rogers, R. R., & Krause, D. W. (2007). Tracking an ancient killer. *Scientific American, 296*(2), 42–51.

Rogers, R. R., Krause, D. W., Rogers, K. C., Rasoamiaramanana, A. H., & Rahantarisoa, L. (2007). Paleoenvironment and paleoecology of *Majungasaurus Crenatissimus* (Theropoda: Abelisauridae) from the Late Cretaceous of Madagascar. *Journal of Vertebrate Paleontology, 27*(S2), 21–31.

Samonds, K. E., Crowley, B. E., Rasolofomanana, T. R. N., Andriambelomanana, M. C., Andrianavalona, H. T., Rami-hangihajason, T. N., Rakotozandry, R., Nomenjanahary, Z. B.,

Irwin, M. T., Wells, N. A., & Godfrey, L. R. (2019). A new late Pleistocene subfossil site (Tsaramody, Sambaina basin, central Madagascar) with implications for the chronology of habitat and megafaunal community change on Madagascar's Central Highlands. *Journal of Quaternary Science, 34*(6), 379–392.

Samonds, K. E., Godfrey, L. R., Ali, J. R., Goodman, S. M., Vences, M., Sutherland, M. R., Irwin, M. T., & Krause, D. W. (2013). Imperfect isolation: Factors and filters shaping Madagascar's extant vertebrate fauna. *PLoS One, 8*(4), e62086. doi: 10.1371/journal.pone.0062086

Scales, I. R. (2014). The Drivers of Deforestation and the Complexity of Land Use in Madagascar. In I. Scales (Ed.), *Conservation and Environmental Management in Madagascar* (pp. 105–125). New York: Routledge.

Schatz, G. (1996). Malagasy/Indo-Australo-Malesian Phytogeographic Connections. In W. R. Lourenco (Ed.), *Biogeographie de Madagascar* (pp. 73–84). Paris: Orstrom.

Schübler, D., Blanco, M. B., Salmona, J., Poelstra, J., Andriambeloson, J. B., Miller, A., Randrianambinina, B., Rasolofoson, D. W., Mantilla–Contreras, J., Chikhi, L., Louis Jr., E. E., Yoder, A. D., & Radespiel, U. (2020). Ecology and morphology of mouse lemurs (*Microcebus* spp.) in a hotspot of microendemism in northeastern Madagascar, with the description of a new species. *American Journal of Primatology, 82*(9), e23180.

Shea, N. (2009). Living on a razor's edge: Madagascar's labyrinth of stone. *National Geographic Magazine, 216*(5), 86–109.

Simpson, G. G. (1952). Probabilities of dispersal in geologic time. *Bulletin of the American Museum of Natural History, 99*(3), 163–176.

Solofondranohatra, C. L., Vorontsova, M. S., Hempson, G. P., Hackel, J., Cable, S., Vololoniaina, J., & Lehmann, C. E. (2020). Fire and grazing determined grasslands of central Madagascar represent ancient assemblages. *Proceedings of the Royal Society B, 287*(1927), 20200598.

Sommer, S. (2003). *Hypogeomys antimena*, Malagasy Giant Jumping Rat,

Vositse, Votsotsa. In S. Goodman & J. Benstead (Eds.), *The Natural History of Madagascar* (pp. 1383–1385). Chicago: University of Chicago Press.

Stearns, S. (1992). *The Evolution of Life Histories*. Oxford: Oxford University Press.

Sterling, E. J. (1993). Patterns of Range Use and Social Organization in Aye-Ayes (*Daubentonia Madagascariensis*) on Nosy Mangabe. In P. M. Kappeler & J. U. Ganzhorn (Eds.), *Lemur Social Systems and Their Ecological Basis* (pp. 1–10). Boston, MA: Springer US.

Sterling, E. J., Betley, E., Sigouin, A., Gomez, A., Toomey, A., Cullman, G., Malone, C., Pekor, A., Arengo, F., Blair, M., Filardi, C., Landrigan, K., & Porzecanski, A. L. (2017). Assessing the evidence for stakeholder engagement in biodiversity conservation. *Biological Conservation, 209*, 159–171.

Stromberg, C. (2011). Evolution of grasses and grassland ecosystems. *Annual Review of Earth and Planetary Sciences, 39*, 517–544. doi: 10.1146/annurev-earth-040809-152402

Stuart, A. J. (2021). Madagascar: Giant Lemurs, Elephant Birds, and Dwarf Hippo. *Vanished Giants* (pp. 176–196). Chicago: University of Chicago Press.

Sussman, R. (1974). Ecological Distinctions of Sympatric Species of *Lemur*. In R. Martin, G. Doyle, & A. Walker (Eds.), *Prosimian Biology* (pp. 75–108). London: Duckworth.

Tattersall, I., & Cuozzo, F. P. (2018). Systématique des Lémuriens Malgaches Actuels (Ordre des Primates) / Systematics of the Extant Malagasy Lemurs (Order Primates). In S. Goodman, M. Raherilalao, & S. Wohlhauser (Eds.), *The Terrestrial Protected Areas of Madagascar: Their History, Description, and Biota* (Vol. 1, pp. 403–421). Antananarivo, Madagascar: Association Vahatra.

Teeling, E. C., Springer, M. S., Madsen, O., Bates, P., O'Brien, S. J., & Murphy, W. J. (2005). A molecular phylogeny for bats illuminates biogeography and the fossil record. *Science, 307*(5709), 580–584.

Thien, L. B., Birkenshaw, C. R., Andriamparany, R., Rabakonandrianina, E., & Schatz, G. E. (2003). Winteraceae, *Takhtajania perrieri*.

In S. Goodman & J. Benstead (Eds.), *The Natural History of Madagascar* (pp. 314–316). Chicago: University of Chicago Press.

Thomas, C. D. (2017). *Inheritors of the Earth: How Nature is Thriving in an Age of Extinction.* New York: PublicAffairs: Hachette Book Group.

Tolley, K., Townsend, T., & Vences, M. (2013). Large-scale phylogeny of chameleons suggests African origins and Eocene diversification. *Proceedings of the Royal Society of London B: Biological Sciences, 280*(1759), 20130184.

Tovondrafale, T., Razakamanana, T., Hiroko, K., & Rasoamiaramanan, A. (2014). Paleoecological analysis of elephant bird (Aepyornithidae) remains from the Late Pleistocene and Holocene formations of southern Madagascar. *Malagasy Nature, 8*, 1–13.

Tsing, A. L. (2015). *The Mushroom at the End of the World: On the Possibility of Life in Capitalist Ruins.* Princeton: Princeton University Press.

Vallet-Coulomb, C., Gasse, F., Robison, L., Ferry, L., Van Campo, E., & Chalié, F. (2006). Hydrological modeling of tropical closed Lake Ihotry (SW Madagascar): Sensitivity analysis and implications for paleohydrological reconstructions over the past 4000 years. *Journal of Hydrology, 331*(1–2), 257–271.

van der Leeuw, S. E. (2000). Land Degradation as a Socionatural Process. In R. J. McIntosh, J. A. Tainter, & S. K. McIntosh (Eds.), *The Way the Wind Blows: Climate, History, and Human Action* (pp. 357–383). New York: Columbia University Press.

Vences, M., Kosuch, J., Rödel, M. O., Lötters, S., Channing, A., Glaw, F., & Böhme, W. (2004). Phylogeography of *Ptychadena mascareniensis* suggests transoceanic dispersal in a widespread African-Malagasy frog lineage. *Journal of Biogeography, 31*(4), 593–601.

Vences, M., & Raselimanana, A. P. (2018). Systématique des Amphibiens Malgaches (Amphibia : Anura) / Systematics of Malagasy Amphibians (Amphibia: Anura). In S. Goodman, M. Raherilalao, & S. Wohlhauser (Eds.), *The Terrestrial Protected Areas of Madagascar: Their History, Description, and Biota* (Vol. 1, pp. 257–288). Antananarivo, Madagascar: Association Vahatra.

Vérin, P. (1980). Cultural Influences and the Contribution of Africa to the Settlement of Madagascar. In UNESCO (Ed.), *Historical Relations across the Indian Ocean: Report and Papers of the Meeting of Experts Organized by UNESCO at Port Louis, Mauritius, from 15 to 19 July 1974* (Vol. 3): UNESCO.

Vérin, P. (1986). *The History of Civilisation in North Madagascar*. Rotterdam: Balkema.

Vieilledent, G., Grinand, C., Rakotomalala, F. A., Ranaivosoa, R., Rakotoarijaona, J-R., Allnutt, T. F., & Achard, F. (2018). Combining global tree cover loss data with historical national forest-cover maps to look at six decades of deforestation and forest fragmentation in Madagascar. *Biological Conservation, 222*, 189–197.

Vieites, D. R., Wollenberg, K. C., Andreone, F., Köhler, J., Glaw, F., & Vences, M. (2009). Vast underestimation of Madagascar's biodiversity evidenced by an integrative amphibian inventory. *Proceedings of the National Academy of Sciences, 106*(20), 8267–8272.

Virah-Sawmy, M., Gillson, L., Gardner, C. J., Anderson, A., Clark, G., & Haberle, S. (2016). A landscape vulnerability framework for identifying integrated conservation and adaptation pathways to climate change: the case of Madagascar's spiny forest. *Landscape Ecology, 31*(3), 637–654.

Virah-Sawmy, M., Gillson, L., & Willis, K. J. (2009). How does spatial heterogeneity influence resilience to climatic changes? Ecological dynamics in southeast Madagascar. *Ecological Monographs, 79*(4), 557–574. doi: 10.1890/08-1210.1

Virah-Sawmy, M., Willis, K., & Gillson, L. (2010). Evidence for drought and forest declines during the recent megafaunal extinctions in Madagascar. *Journal of Biogeography, 37*(3), 506–519. doi: 10.1111/j.1365-2699.2009.02203.x

Voarintsoa, N. R. G., Railsback, L. B., Brook, G. A., Wang, L., Kathayat, G., Cheng, H., Li, X., Edwards, R. L., Rakotondrazafy, A. F. M., & Razanatseheno, M. O. M. (2017). Three distinct Holocene intervals of stalagmite deposition and nondeposition revealed in NW Madagascar, and their paleoclimate implications. *Climate of the Past*, 13(12), 1771–1790.

Vorontsova, M. S., Besnard, G., Forest, F., Malakasi, P., Moat, J., Clayton, W. D., Ficinski, P., Savva, G. M., Nanjarisoa, O. P., Razanatsoa, J., Randriatsara, F. O., Kimeu, J. M., Luke, W. R. Q., Kayombo, C., & Linder, H. P. (2016). Madagascar's grasses and grasslands: Anthropogenic or natural? *Proceedings of the Royal Society B, 283*(1823), 20152262.

Wake, D. B., & Vredenburg, V. T. (2008). Are we in the midst of the sixth mass extinction? A view from the world of amphibians. *Proceedings of the National Academy of Sciences, 105*(Supplement 1), 11466–11473.

Wang, L., Brook, G. A., Burney, D. A., Voarintsoa, N. R. G., Liang, F., Cheng, H., & Edwards, R. L. (2019). The African Humid Period, rapid climate change events, the timing of human colonization, and megafaunal extinctions in Madagascar during the Holocene: Evidence from a 2m Anjohibe Cave stalagmite. *Quaternary Science Reviews, 210*, 136–153.

Watson, R. T. (2018). Raptor Conservation in Practice. In J. Sarasola, J. Grande, & J. Negro (Eds.), *Birds of Prey* (pp. 473–498). Springer International.

Watson, R. T., de Roland, L. A. R., Rabearivony, J., & Thorstrom, R. (2007). Community-based wetland conservation protects endangered species in Madagascar: Lessons from science and conservation. *BANWA Archives (2004–2013), 4*(1), 83–97.

Watson, R. T., & Rabarisoa, R. (2000). Sakalava fishermen and Madagascar Fish Eagles: Enhancing traditional conservation rules to control resource abuse that threatens a key breeding area for an endangered eagle. *Ostrich, 71*(1–2), 2–10.

Wegener, A. (1929). *The Origin of Continents and Oceans. [English translation (1966), of the 4th German edition, by J. Biram.]* New York: Dover Publishing.

Wells, H. G. (1894). *Aepyornis Island.* London: The Pall Mall Budget.

Wells, N. (2003). Some Hypotheses on the Mesozoic and Cenozoic Paleoenvironmental History of Madagascar. In S. Goodman & J. Benstead (Eds.), *The Natural History of Madagascar* (pp. 16–34). Chicago: University of Chicago Press.

Wells, N. A., & Andriamihaja, B. (1993). The initiation and growth of gullies in Madagascar: Are humans to blame? *Geomorphology, 8*(1), 1–46. doi: Doi 10.1016/0169-555x(93)90002-J

Wells, N. A., Andriamihaja, B., & Rakotovololona, H. F. S. (1991). Patterns of development of lavaka, Madagascar's unusual gullies. *Earth Surface Processes and Landforms, 16*(3), 189–206. doi: DOI 10.1002/esp.3290160302

Western, D., & Maitumo, D. (2004). Woodland loss and restoration in a savanna park: A 20-year experiment. *African Journal of Ecology, 42*(2), 111–121.

White, F. (1983). *The Vegetation of Africa: A Descriptive Memoir to Accompany the UNESCO/AETFAT/UNSO Vegetation Map of Africa* (Vol. 20). Paris: UNESCO.

Willis, K., Bailey, R., Bhagwat, S., & Birks, H. (2010). Biodiversity baselines, thresholds and resilience: Testing predictions and assumptions using palaeoecological data. *Trends in Ecology & Evolution, 25*(10), 583–591.

Wilmé, L., Goodman, S. M., & Ganzhorn, J. U. (2006). Biogeographic evolution of Madagascar's microendemic biota. *Science, 312*(5776), 1063–1065.

Wilmé, L., Waeber, P. O., & Ganzhorn, J. U. (2017). Human translocation as an alternative hypothesis to explain the presence of giant tortoises on remote islands in the south-western Indian Ocean. *Journal of Biogeography, 44*(1), 1–7.

Wright, H. T., & Rakotoarisoa, J-A. (1997). Cultural Transformations and Their Impacts on the Environments of Madagascar. In S. Goodman & B. Patterson (Eds.), *Natural Change and Human Impact in Madagascar* (pp. 309–330). Washington, DC: Smithsonian Institution Press.

Wright, H. T., & Rakotoarisoa, J-A. (in press). The Rise of Malagasy Societies: New Developments in the Archaeology of Madagascar. In S. M. Goodman (Ed.), *The New Natural History of Madagascar* (2nd ed.). Princeton: Princeton University Press.

Wright, H. T., Vérin, P., Ramilisonina, Burney, D., Burney, L. P., &

Matsumoto, K. (1996). The evolution of settlement systems in the Bay of Boeny and the Mahavavy River Valley, north-western Madagascar. *Azania: Journal of the British Institute in Eastern Africa, 31*(1), 37–73.

Wright, P. C. (1999). Lemur traits and Madagascar ecology: Coping with an island environment. *American Journal of Physical Anthropology, 110*(S29), 31–72.

Wright, P. C., & Andriamihaja, B. (2002). Making a Rain Forest National Park Work in Madagascar: Ranomafana National Park and Its Long-Term Research Commitment. In J. Terborgh, C. van Schaik, M. Rao, & L. Davenport (Eds.), *Making Parks Work: Strategies for Preserving Tropical Nature* (pp. 112–136). Covelo, CA: Island Press.

Wright, P. C., Tecot, S. R., Erhart, E. M., Baden, A. L., King, S. J., & Grassi, C. (2011). Frugivory in four sympatric lemurs: Implications for the future of Madagascar's forests. *American Journal of Primatology, 73*(6), 585–602.

Wright, R., & Askin, R. (1987). The Permian–Triassic boundary in the southern Morondava Basin of Madagascar as defined by plant microfossils. *Gondwana Six: Stratigraphy, Sedimentology, and Paleontology, 41*, 157–166.

Yamashita, N., Tan, C. L., Vinyard, C. J., & Williams, C. (2010). Semi-quantitative tests of cyanide in foods and excreta of three *Hapalemur* species in Madagascar. *American Journal of Primatology, 72*(1), 56–61.

Yang, Z., & Yoder, A. D. (2003). Comparison of likelihood and Bayesian methods for estimating divergence times using multiple gene loci and calibration points, with application to a radiation of cute-looking mouse lemur species. *Systematic Biology, 52*(5), 705–716.

Yoder, A. D., Burns, M. M., Zehr, S., Delefosse, T., Veron, G., Goodman, S. M., & Flynn, J. J. (2003). Single origin of Malagasy Carnivora from an African ancestor. *Nature, 421*(6924), 734.

Yoder, A. D., Campbell, C. R., Blanco, M. B., dos Reis, M., Ganzhorn, J. U., Goodman, S. M., Hunnicutt, K. E., Larsen, P. A., Kappeler, P. M., Rasoloarison, R. M., Ralison, J. M., Swofford, D. L., & Weisrock, D. W. (2016). Geogenetic patterns in mouse lemurs (genus

Microcebus) reveal the ghosts of Madagascar's forests past. *Proceedings of the National Academy of Sciences, 113*(29), 8049–8056. doi: 10.1073/pnas.1601081113

Yoder, A. D., & Nowak, M. D. (2006). Has vicariance or dispersal been the predominant biogeographic force in Madagascar? Only time will tell. *Annual Review of Ecology, Evolution, and Systematics, 37,* 405–431.

Yoder, A. D., & Yang, Z. (2004). Divergence dates for Malagasy lemurs estimated from multiple gene loci: Geological and evolutionary context. *Molecular Ecology, 13*(4), 757–773.

Yonezawa, T., Segawa, T., Mori, H., Campos, P. F., Hongoh, Y., Endo, H., Akiyoshi, A., Kohno, N., Nishida, S., Wu, J., Jin, H., Adachi, J., Kishino, H., Kurokawa, K., Nogi, Y., Tanabe, H., Mukoyama, H., Yoshida, K., Rasoamiaramanana, A., Yamagishi, S., Hayashi, Y., Yoshida, A., Koike, H., Akishinonomiya, F., Willerslev, E., & Hasegawa, M. (2017). Phylogenomics and morphology of extinct paleognaths reveal the origin and evolution of the ratites. *Current Biology, 27*(1), 68–77.

INDEX

Page references in *italics* indicate images.

aardvark 85
Aepyornis spp. See elephant birds
aerial photography 218–9
Africa
    climate 39-40, 112
    continental drift 27, 29-31
    grassland animals 104, 120,
        88, 137-9, 204-6
    Miombo woodland 140-1
    national parks 104
    ocean currents 32-3
    shared plant ancestry 99-100,
        106, 109
Afrosoricida 83
agriculture
    agroforestry *See* Imorona
    field preparation 179
    fire and 17, 94, 110–1, 142,
        182-3
    earliest evidence 156-7
    history 163-70, 185-93, 216-7
    maize 38, 224-5
    rainfall impact on 38, 76
    rice *See* rice

    shifting or swidden cultivation
        217, *colour fig 25*
    soil erosion and 43
    vanilla *See* Imorona
Amboseli National Park *See*
    Africa, national parks
Amphibians *69, 72, 221 See also*
    ancient plants and animals
    and frogs
Ampasambazimba sub-fossil site
    See central highlands
Ampijoroa forest *19*, 65, 79, 96
Analafaly village 37–8, 239–44
ancient plants and animals
    50-61
Andohahela National Park *See*
    national parks and nature
    reserves
Andranosoa early settlement site
    164–5, 171
Andriamamelo Cave 202–4
Andrianampoinimerina, King
    186, 227
Andringitra mountains 35–6

331